PROFITABLE FREE RANGE
EGG PRODUCTION

PROFITABLE FREE RANGE EGG PRODUCTION

MICK DENNETT

The Crowood Press

First published in 1995 by
The Crowood Press Ltd
Ramsbury, Marlborough
Wiltshire SN8 2HR

British Library Cataloguing-in-Publication Data
A catalogue record for this book is available from the British
Library.

ISBN 1 85223 835 6

Picture Credits
Line-drawings by Claire Upsdale-Jones
All photographs by the author, except those on pages 20
(bottom), 27, 39, 75 and 76, which are reproduced by courtesy
of Willow Agriculture.

Acknowledgements
Peter Barton for editing this book and writing the chapter on
marketing; Kim Miller of MAFF for writing the section on egg
packing station regulations; Bruce Pattern of MAFF for his help
and advice on marketing regulations; Dr Tony Marangos BSc,
CBiol, MiBiol, for his technical assistance; Abbott Laboratories
US and UK for their technical help with the drawings; David
Hall of East Sussex County Council for his help on planning
regulations; ISA Poultry for the various graphs and relevant
information; Richard and Sue Newble for drawings and
photographs; Broiler Equipment Company and Plasson
Drinkers (Anglian Livestock Appliances) for drawings; to Willow
AG for the use of photographs; and all the farm owners who
allowed me to take photographs for publication in this book: P.
Cook, M. Hanbury, S. Jones, D. Litchfield, T. Parsons, R. Hall, R.
Cox, J. Durham, P. Barton, D. Rusbridge and S. Laughlin.
Thank you to my patient wife Angela.

Typeset by Intype, London
Printed and bound in Great Britain by The Bath Press

Contents

Introduction

The poultry industry has seen many changes since one of the common ancestors of the domestic fowl, the wild red jungle fowl of India called *Gallus bankiva*, provided meat and eggs for the local inhabitants. Selective breeding from this bird, and many others from all over the world, produced excellent dual-purpose breeds like the Light Sussex and Wyandotte. The females were very good layers and both the males and females made good table birds. Since the 1960s the breeders have developed strains from these old breeds which contain certain defined characteristics. Hence today there are no commercial dual-purpose birds, as the breeders have separated the egg-laying strains from the meat-producing strains to supply two very different and specialized markets.

Until the first half of the twentieth century the majority of eggs were produced using extensive free range systems. These systems included the fold type of movable house which was a small wooden house with a run attached to hold up to 25 layers. The smallest type of house in use was the Sussex Ark, which was about 6ft long by 3ft wide (180 x 90cm), holding up to twelve layers. Semi-intensive houses, such as the straw house, straw yard and larger wooden, specially designed, fixed houses, were used from the 1920s onwards, but owners were not keen on static houses, mainly because of disease problems and the uneven distribution of manure which did not fit in with the crop rotation on the land.

In 1947 there were 22,000 laying stock scattered all over the British Isles, of which 85 per cent were kept in flock sizes of one to 100 birds. In the 1940s drugs and vaccines, used to control coccidiosis and viral diseases, were not very successful, which is probably the reason why, in 1947, 2,222 cases of fowl pest were confirmed. The average mortality of all the national laying test in the 1940s was approximately 12 per cent, so the old extensive free range systems were very labour-intensive and suffered from a high mortality rate.

The introduction of the battery cage from the USA just after the Second World War reduced both mortality and labour costs, with stocking densities as high as 20,000 birds to the acre (49,420 birds per hectare). It is not surprising that by the 1960s, the battery system was responsible for 95 per cent of all commercial egg production. Today, the figure is nearer 84 per cent. Not all of the good laying results were due solely to the battery system, however. Great improvements were being made by the breeders in the 1960s to increase the amount of eggs each bird would lay and to reduce the amount of feed eaten. Modern hybrid hens now lay nearly twice as many eggs as their predecessors.

Since the mid-1980s there has been a gradual resurgence of the extensive free range egg producing systems, but today they are not as labour-intensive and the results are more predictable. This about-turn in the way eggs are produced is due to the changing attitudes of the public towards animal welfare, which has led to the development of a strong animal welfare lobby, both in Europe and world-wide. Increasing concern is being shown in the welfare of the animals which produce our food. In this environment of changing attitudes it is important for the commercial egg producer to try to adapt his system of egg production in a way that will reflect the feelings of the general public.

Great Britain leads the world in developing modern, efficient and welfare-conscious alternatives to the battery cage. No system of agriculture is, however, perfect. Free range systems give birds freedom to follow all their natural characteristics and instincts, both good and bad. It is the bad points that make keeping a large amount of birds in a relatively small area difficult. If the bird management is not absolutely correct in a battery system it will have little or no effect on the birds' welfare and egg production. If the same degree of mismanagement occurs in a free range system, both egg production and bird welfare will be adversely affected. The systems described in this book will produce good results, as long as you are aware of the necessity to provide the birds with a very high standard of stock management all day and every day. Provided you do not mind working seven days a week applying a high standard of farm management, nor expect to become rich overnight, a reasonable standard of living should be achieved, together with the satisfaction of being your own boss.

— 1 —

Know Your Market

No apology is made for commencing a book on egg production with a chapter on marketing. Armed with this book and a commitment to pay attention to detail, eggs will undoubtedly be produced, but as we are examining commercial egg production it is important that the end result is not just eggs but profits, which are the result of selling a product at greater than the cost of production. This simplistic approach is important to differentiate between the hobby producer and the commercial producer; size is not the essence, rather the attitude of mind. Many producers start out well intentioned, but lose sight of the fact that without a secure market for their product the project is doomed, and every year hundreds of operations fail after huge personal loss, including the family farm and even the home.

MARKET RESEARCH

Your first step should be to research your intended market. For a small operation that will mean talking to prospective customers, including perhaps the local shops, school or public house. Discover who their existing suppliers are and what prices they are selling for. Are they already supplied with free range eggs and, if so, are they happy with the service and quality they are receiving? Larger production units will need to explore markets with national or local egg packers. Members of the National Farmers' Union can obtain valuable market information from their regional poultry specialist, as well as from ADAS, the Agricultural Development and Advisory Service (*see* Useful Addresses).

THE BUSINESS PLAN

Once the research has been completed it is essential to draw up a detailed budget and costing, only then can anticipated profit margins be calculated. Estimated costs of production are invariably too low and expected profits too high; anticipated profits should therefore be halved to give a reasonable margin for safety. If after completing these calculations the enterprise still stands up, it is likely to succeed.

Whereas commercial industries outside agriculture aim to achieve a return on capital of, say, 10 per cent, agriculture has tended to achieve a mere 2 per cent. Whilst this might seem a paltry sum, there are other factors, such as increased land values, the possibility of gaining planning permission for a home on a green-field site, plus of course the pleasure of living and working on your own enterprise, to take into account.

MARKET OPPORTUNITY

Producing for a Packer

Any free range unit with more than 3,000 birds will, in all probability, be looking to an egg packer as an outlet for its egg production. In this case the packer is your customer and will be looking for both service and, above all, quality, as the eggs will predominantly be sold to the major multiples which are national names. Packers have exacting quality standards covering: freshness and refrigeration; the specification of feed used; internal egg quality; welfare conditions; shell colour and strength; egg hygiene; and farm hygiene. The multiples require these standards from their packers who, in turn, require them from their suppliers. Much of the quality auditing is now done by the packers and they undertake regular inspections of their free range producers to see that the specifications are being adhered to. Failure can mean instant loss of contract and financial disaster.

Advantages
1. You will have a regular commitment to take all the eggs from the farm.
2. The packer will take all the eggs, even unpopular sizes as well as second-quality eggs.
3. You will only have one customer to handle.
4. All transport and packaging is normally provided.
5. You will incur no capital cost to provide grading and packing equipment or staff.

Disadvantages

1. Producer prices are fixed by the packer with little relevance to the cost of production and often with little negotiation.
2. Management standards must be higher and so costs are higher.

Direct Selling

It tends to be true that to be successful in free range production you have to be either small or large. The smaller unit usually concentrates on selling direct to local shops, public houses, factories, offices, estates, and even to local housing estates, or via the farm gate or farm shop. Selling at the farm gate to the final consumer does not require registration as a packer, with all the incumbent equipment and regulation this requires.

Larger units with multi-aged flocks have traded successfully as independent packers supplying a large range of independent shops. However, very few have been able to meet the exacting standards required by the national supermarket chains.

Advantages

1. The ability to negotiate deals direct with your customer.
2. Prices tend to be a truer reflection of the state of the market.
3. Quality standards tend to be a little lower and returns per bird therefore marginally higher.

Disadvantages

1. You must have multi-aged flocks to ensure continuity of supply.
2. It is sometimes difficult to market smaller sizes.
3. Capital costs and staffing for packing and grading can be high.
4. Delivery and packaging costs are high.

Franchising

Joining a free range franchise can be a speedy way into the business. A franchiser will often offer a complete package, including building, equipment, pullets and feed plus training and a contract to purchase all the eggs produced. As with any business opportunity, you must be careful to investigate all aspects, including contract terms and prices. Convenience may have an added cost which can be considerable. For example, a £20 per tonne difference in the cost of your feed can mean £0.04 increase in the cost of production. This in turn equates to £0.85 per bird loss of profit, or for an average flock of, say, 3,000 birds a loss of £2,550 in income.

Advantages
1. The convenience of an off-the-shelf integrated product.

Disadvantages
1. A lack of control on major cost inputs.

THE STRUCTURE OF THE EGG INDUSTRY

The egg market, like that of other fresh foods, is constantly evolving to meet new demands. For example, in the 1970s consumers associated freshness with buying directly from a local supplier or corner shop. Today the majority of eggs are purchased from supermarkets who, to their credit, have established an enviable standard for quality. Today's problem is not so much one of freshness but of education. Complaints, such as the inability to peel a hard-boiled egg because it is stale, illustrate public ignorance. Many eggs are now on the shelf a day after they were laid, and an egg that will not peel easily is likely to be less than one week old and it is therefore a sign of peak freshness!

The movement towards one-stop shopping has had a major impact on egg production and this, coupled with a steady decline in shell egg consumption, has led to a reduction in the market for independent producers selling eggs direct. The accelerating closure of corner shops has had a dramatic effect on battery producers, but the future for free range producers is less severe as there remains an increasing demand for free range eggs and consumers are known to be willing to travel some distance to buy their eggs direct.

SUPPLYING MAJOR MULTIPLE OUTLETS

Although the national supermarket chains profess an interest in buying eggs direct from local suppliers, their logistical demands make it uneconomical in practice. Central distribution is now used by all major retailers and this requires the mandatory use of their 'own label', requiring a huge investment in meeting their demands for packaging and labelling and putting a 'best before' date on each egg. The cost of the egg printer alone would be prohibitive to all but the largest egg packers.

Few independent free range producers have cleared all the hurdles to supply the multiples and fewer still have continued supplying direct. It is for this reason that the large packers, with their fleets of lorries and packing stations in each region, exclusively serve this growing area of the market which is expected to reach 75 per cent of total sales by the end of the twentieth century.

THE FUTURE OF THE INDEPENDENT PRODUCER

There will always be a demand for eggs from local independent stores, although those that remain are likely to form buying groups to purchase competitively. The national egg packers regard their core business as supplying the multiples, but it seems inevitable that, with their huge capacity, they will sell to these buying groups. The egg market in the UK has always been very competitive and the humble egg regarded simply as a commodity to be traded. The business is extremely cyclical, based as it is upon supply and demand. One good year is inevitably followed by two poor years with prices often at or below the cost of production until sufficient birds are removed and supply is once more balanced with demand. The ease with which traders can import eggs from France and Holland adds to the problem, because a shortage of eggs in the UK and an expectation of improved prices can be quickly offset if Europe has a surplus, which is then transferred to the UK.

Fortunately, the free range market is currently immune from such considerations as the rest of Europe has no market for the product and therefore has no available production to expand into the UK. Few national egg packers have gone into free range production themselves, choosing instead to take supplies from independent producers. This might seem perverse in a market dominated by efficiency and driven by profit. The fact that the investment and capital demand of a free range unit is so high makes it difficult for them to contemplate production at present. Take, for example, the fact that with the purchase of land a new free range production unit might incur twice the cost per bird of that housed in a battery unit. A modest 50,000 bird farm built to free range specification would cost in the region of £750,000 and require management who are prepared to work late at night, perhaps to lock up birds at ten o'clock in the evening. An independent, self-employed farmer is used to working such hours, but finding employees willing to work a fourteen-hour day is not easy.

— 2 —

Getting Started

SELECTING THE LAND

The search for ideal land for a free range production unit is of the utmost importance. Light, well-drained, sheltered land will support hens for many years, whilst heavy, waterlogged land should be avoided at all costs. It is also important to check that there is good access to roads and mains services, including electricity and water, and that there are no local by-laws prohibiting the keeping of poultry in the locality or unusual restrictions on building height.

It is possible to keep up to 400 birds per acre (990 birds per hectare) on well-established pasture and birds will thrive in wood-

Trees will encourage birds away from the poultry house.

Shrubs and trees provide the birds with security and shade.

land as they enjoy being out of the direct glare of the sun. Trees situated within 100yd (110m) of the poultry house will encourage the birds to range and give them a feeling of security, and shrubs and hedgerows also provide shade. Provided birds are not let out of the house too early each morning they will not lay their eggs in bushes or hedgerows.

MIXING POULTRY WITH OTHER STOCK

On many farms the outer areas of grassland will not be grazed well enough to keep the grass short. Sheep, horses and goats can all be used to keep the grass short, provided that proper fencing is put up to keep them away from the poultry house. Failure to do this could lead to severe stress and smothering should the birds be frightened.

CHOOSING THE RIGHT BIRD

It is important to remember that the modern hybrids used in the

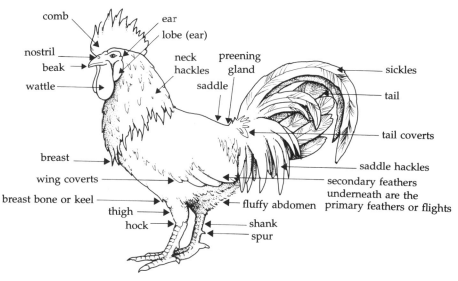

The main points of the fowl.

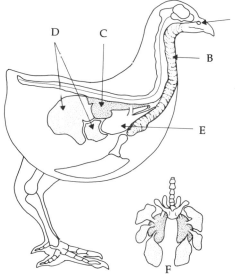

A   nostril

B   trachea

C   lung

D   air sacs

E   heart

F   air sacs and lungs

The respiratory system of the fowl.

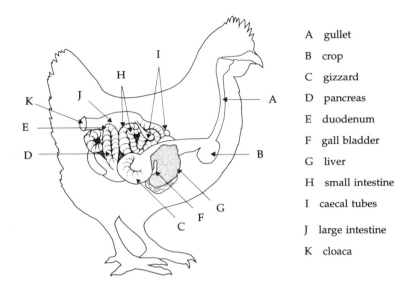

A	gullet
B	crop
C	gizzard
D	pancreas
E	duodenum
F	gall bladder
G	liver
H	small intestine
I	caecal tubes
J	large intestine
K	cloaca

The digestive system of the fowl.

poultry industry are primarily bred for a life in a battery cage. However, good results can also be obtained when used on alternative systems. It is probable that the wide variations in production that have been found to occur on free range units is more attributable to management problems than breed. Indeed, many free range units have consistently beaten the breeder's targets for egg numbers, egg size and mortality.

There are several different hybrids which are being used successfully on barn, free range and perchery systems of egg production. The two most widely used are the ISA Brown and the Hisex Brown, which have both performed very well over many years, producing over 300 eggs in 52 weeks' lay. Not only do both breeds produce plenty of eggs, but the quality and size of the eggs is very good. In more recent years other breeds have been introduced, including the Lohmann Brown, Hy-Line, Shaver 579 and Black Rock, which are performing well in barn, free range and perchery systems.

The graph opposite gives the results of an actual free range flock and so shows the standards that can be expected from the ISA Brown hybrid, assuming that the management of the bird is correct. The method used for calculating the various production and feeding information from a flock of birds can be found on pages 117–18.

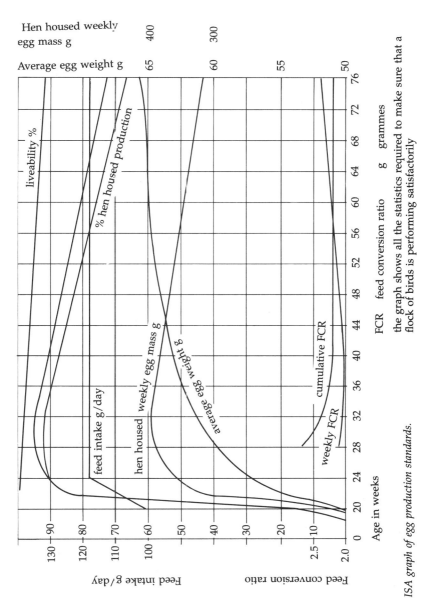

ISA graph of egg production standards.

17

REGISTRATION AS AN EGG PACKING STATION

The British Egg Marketing Regulations (available from MAFF Egg Inspectorate Division) do not apply if eggs are being sold directly to consumers from your own premises, a stall at a car boot sale or public market (with the exception of auction markets) or by door-to-door selling. However, if you are supplying shops it is necessary to register as an egg packing station with the Ministry of Agriculture's Egg Marketing Inspectorate. This includes sales to farm shops (other than on your own premises), garden centres and garages. Producers outside Great Britain should consult their own authorities for regulations which govern them.

To qualify for registration, premises must meet certain requirements.

1. They must be sufficiently large to accommodate the volume of work to be done.
2. They must have adequate ventilation and lighting, be able to be cleaned and disinfected, and be able to prevent the eggs from being affected by excessive heat and cold. This may mean taking steps such as cladding the walls or constructing a false ceiling. It is important that any windows are covered to prevent direct sunlight entering the building and also to help when candling the eggs. The floor should be of a type which is easy to keep clean.

An egg grading room in the end of the poultry house.

Weighing the eggs weekly to find the average egg weight.

3. The premises must be used only for handling eggs, or for the storage of other products which will not affect the eggs. Eggs can become tainted from some products.

The following equipment is also required:

1. A machine for grading the eggs into the eight weight grades, which range from Size 0 (75g and over) down to Size 7 (under 45g), with a 5g band for each grade in between. The type of machine required will depend on the throughput of eggs. Machines are available which will grade from 1,600 eggs per hour up to hundreds of cases of eggs per hour. Even the smallest grader would be very expensive new, but second-hand ones are available. If throughput is very small, it is possible to use a suitable pair of scales to grade the eggs and a candling lamp to extract faults. Candling lamps are built into egg grading machines.
2. A device for measuring the air cell in the egg.
3. If a grading machine is used a set of scales for checking the grading machine's accuracy will also be required.

To register as an egg packing station, contact the Regional Egg

A battery-powered egg collecting trolley which holds 2,000 eggs.

An egg trolley in the egg store.

Marketing Inspector for your area (or the equivalent outside the UK), who will be able to give more detailed information on registration and send relevant leaflets and an application form. When your application is received, it will be acknowledged and a pre-arranged visit will be made by an Egg Marketing Inspector within fourteen working days. If the premises and equipment comply, you will receive a letter giving the packing station a registration number. If they do not, you will be issued with a letter listing the equipment needed and work required in order to meet the standard and a suitable time limit will be given within which to complete the necessary work.

PLANNING PERMISSION

The Town and Country Planning General Development Order 1988 grants planning permission for a wide range of developments associated with agricultural uses of land on units of $12^1/_2$ acres (5 hectares) or more. In certain cases this planning permission cannot be exercised unless the farmer has applied to the Local Planning Authority in whose area the proposed development is to take place. Your Local Planning Authority (or equivalent outside the UK) should therefore be approached in the first instance.

These regulations also apply to significant extensions and alterations to existing buildings and roads, and so on. New or old buildings to be used for the accommodation of livestock or associated structures, such as slurry tanks or lagoons, which are built within 440yd (400m) of the curtilage of a protected building (i.e. private house) must have full planning permission.

Full planning permission is not required for poultry houses provided they are mobile or portable units. Some Local Authorities do not regard houses that are connected to mains water and electricity as mobile and portable and these therefore will require planning approval. Avoiding this often entails taking water to the birds daily and using batteries to give birds enough light in the winter. Small mobile houses are very labour-intensive and one person would do well to manage 2,000 layers without any help. Therefore think carefully before investing in a large mobile commercial laying enterprise.

Living on the Farm
There are few opportunities to obtain planning permission for a new

21

house in the countryside, with most Local Authorities operating a near total prohibition against new rural dwellings. The need for an agricultural worker or farmer to live on or close to his farming enterprise is, however, one of the few exceptions and, subject to establishing a clear agricultural need, many free range producers have successfully obtained planning permission to erect a dwelling on their land.

Local Planning Authorities can, where they think it appropriate, ask for either a functional or financial test (or both) to justify evidence of the genuineness of someone's intentions regarding a new enterprise. In cases where the evidence supporting an application for an agricultural dwelling is inconclusive, perhaps because there is uncertainty about the viability of the proposed enterprise, the Local Planning Authority will consider permission for a caravan or other temporary accommodation. When such permission is granted it will usually have to be renewed after a set number of years.

3

Housing

SITING THE POULTRY HOUSE

The house described in this book can be used for any of the three systems of commercial egg production, barn, free range or deep litter. Only the internal design needs to be altered to suit whichever system of production is chosen. Externally the house has been designed to be 'visually friendly'. It is very important to consider the visual effect of a building in the landscape and to adopt a positive approach towards the opinions of your neighbours and local council.

Avoid building a large poultry house on the top of a hill as this would do little to enhance the beauty of the countryside and would be particularly vulnerable to critical public opinion. On a practical level, the house would be exposed to strong cross-winds in the winter, making it cold and draughty, which would increase feed costs unnecessarily as the birds would consume extra feed to keep warm.

Do not put the house at the bottom of a valley as the natural flow of the water collecting around it, combined with thousands of scratching feet, would turn the land into a mud bath. Dirty feet would increase the number of dirty eggs, which would not only reduce income as these eggs would be second-grade, but would also make birds more susceptible to disease.

The ideal position would be half-way up a south-facing field, with a slight gradient taking away any rain water past the house to the lower levels. Avoid clay-type soil which does not drain well; sandy or loam-type soils make the best grazing fields for free range hens.

Whatever the type of soil, it must be well drained. A French drain should be laid all round the house approximately 2yd (2m) away from the walls, and the 2in (5cm) beach filling should not only fill the trench but be extended to provide a clean apron all around the house. Shrubs and trees a short distance from the house

A centre passageway house showing the French drain going away from the corner of the house.

will encourage birds to move away from the pop holes and provide shade in the summer months.

HOUSE DESIGN

The house design and dimensions shown in the diagrams on pages 25 and 26 produce excellent results, provided their management is satisfactory. It has been designed to be visually acceptable as well as functional. Try to visualize the new development in relation to any existing buildings on the farm and see if all the buildings blend in together. An overhanging roof will provide a shadow to cool the incoming air and reduce light intensity in the summer months, as well as adding a little character to the shape of the house.

Dimensions

The dimensions of the house given in the diagrams are critical and should not be altered. However, if a greater number of birds is required, it is possible to extend the length of the house and add pens or divisions pro rata. If the distance between the nest boxes and the pop holes or outer wall is greater than 6.7yd (6.1m), there is an increased risk of floor eggs as the birds lose sight of the nest boxes. The high roof is important as this will keep the warm used

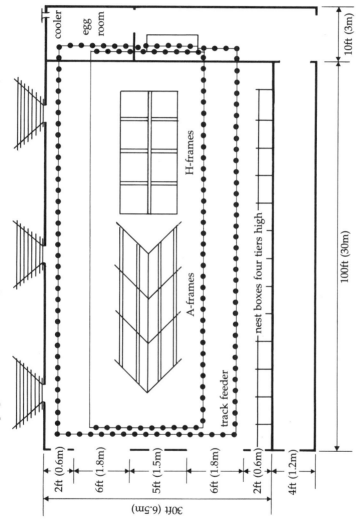

Plan of a poultry house with a side egg collection passageway.

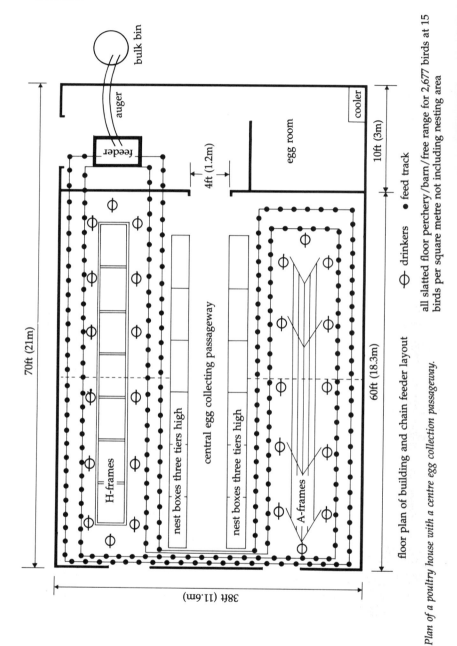

bulk bin

auger

feeder

4ft (1.2m)

egg room

cooler

10ft (3m)

70ft (21m)

H-frames

nest boxes three tiers high

central egg collecting passageway

nest boxes three tiers high

A-frames

60ft (18.3m)

38ft (11.6m)

⊕ drinkers   • feed track

all slatted floor perchery/barn/free range for 2,677 birds at 15 birds per square metre not including nesting area

floor plan of building and chain feeder layout

Plan of a poultry house with a centre egg collection passageway.

air well above the birds' heads. This is of particular importance when there is a power failure and natural air convection must be relied upon to keep the air fresh.

Divisions or pens prevent birds drifting from one end of the house to the other. Without these restrictions some of the equipment will become greatly over-used, causing fierce competition for food and water amongst the birds (*see* Chapter 5). Also, the temperature within the house will vary considerably as the end of the house with fewer birds becomes cooler and the overstocked end becomes warmer, making it impossible to keep an even temperature all over the house. The windows or air inlets must be high over the birds' backs to avoid cold draughts blowing on to them, as cold, draughty houses will cause an uneven distribution of birds on the floor as the birds move to the warmer parts of the house.

BUILDING MATERIALS

The building can be constructed of a steel or wooden framework.

A laying house with sun veranda or scratching area.

Floor

The floor should be made of concrete. Chalk can be used, but cement powder should be added until a grey colour is obtained in the 'floated-up' wet chalk as this will create a smooth easy-to-clean surface. Earth floors are difficult to clean out and wash down which can make disease eradication almost impossible.

Deep Pit and Passageway Walls

The external walls of the deep pit and internal walls of the passageways should be built with heavy 6in (15cm) concrete blocks as these blocks will be supporting other heavy structures in the house. None of the deep pit's contents should be allowed to contaminate the egg collecting passageways or any other clean area.

Walls and Roof

These can be made of wood or any of the lightweight metal claddings available. Try to choose colours that harmonize the building into the surrounding landscape. Whatever type of materials are used, the walls and roof must be insulated, otherwise the wide variation in temperatures will cause condensation to form and make the house cold and wet in the winter. Poor insulation will also cause high temperatures in the summer, often leading to high levels of ammonia in the house which will affect the birds' health and internal egg quality. The ammonia will also be offensive to the staff working in the house.

Slatted Floors and Roosting Pits

From personal experience, wooden and plastic slats create less floor egg problems than wire floors. The slats covering the deep pit on the all slatted system and the roosting area in the deep litter system should be made of hardwood 1in (25mm) square with a cambered top edge. There must be a 1in (25mm) space between the slats; if this distance is any greater the birds will get their heads stuck in the gap. Heavy-gauge wire can be used but will sometimes bend or sink, making the ideal place for a hen to lay her eggs. Plastic slats are also very good and have the advantage of being smaller and lighter than wood or wire. Do not make the slats too big or you will have great difficulty getting them out of the house at cleaning-out time, although the slats can be hinged to the pit struc-

A side or centre passageway house.

An overhanging roof adds character and shade to a poultry house.

29

tures allowing the cleaning-out to take place without taking the slats outside. One of the cheapest materials to use for slats is large size slate battening, but check for any weak knots in the wood. The heavy concrete block pit walls will provide the support required for the main weight-bearing timbers the slats will rest upon.

It is very helpful to paint all wooden surfaces with creosote to reduce the risk of red mite establishing themselves in the house. This needs to be done at least one month before the birds come in.

For a stocking density of 15 birds per sq. metre (12.5 birds per sq. yard) the deep pit will need to be $1^1/_4$yd (1.2m) deep. The pit will have to be 12in (30cm) deeper if it is intended to force-moult the birds or stock at 20 birds per sq. metre (16.7 birds per sq. yard) (*see* forced moulting, page 88).

AIR INLETS OR WINDOWS

The recommendations given below are made on the assumption that extraction fans will be placed on the roof, along the top ridge of the house, allowing natural air convection to take place during a power failure. The amount of air inlets or windows required in a house will vary according to the number of birds and the type of ventilation system used. Houses stocked at more than 15 birds per sq. metre (12.5 birds per sq. yard) must have fully automatic ventilation, including a generator. A fully automatic ventilation system, using fans and thermostats with a generator, will only require sufficient air inlets to allow the fans to operate at maximum air extraction. The fan manufacturers will provide information on the minimum and maximum air inlet requirements for the fans required.

If the house is not equipped with a generator, it is very important that it has sufficient air inlet space to provide adequate air and light when there is a power failure. To work out how much air inlet space is required for an emergency situation, it is necessary to know the amount of floor space available to the birds. From the house on page 25 it can be seen that the total space available to the birds is 2,040ft² (190m²). The window space required is 10 per cent of the total floor space available to the birds, i.e. 204ft² (19m²) or 102ft² (9.5m²) of window space placed equally down each side of the house. Six solid (not glass) windows measuring 7ft x $2^1/_2$ft (210 x 76cm) placed equally down each side of the house (12 windows in total) will give the minimum air inlet space required in an emer-

An enclosed slatted ramp stops the birds getting underneath.

gency for a house stocked at 15 birds per sq. metre (12.5 birds per sq. yard). The pop holes will give extra air space if required.

If the stocking density is higher than 15 birds per sq. metre (12.5 birds per sq. yard), a generator will be necessary to ensure the birds have sufficient ventilation to keep them happy and healthy. All the windows or vents must be baffled from the light and wind on the outside, but the baffles or hoods must be portable for quick removal in an emergency. Ensure that the internally opening windows are baffled at the sides to prevent any cold air falling on the birds' backs. Check that the cross beams in the roof do not direct the cold incoming air on to the birds' backs.

It is important to ensure the birds cannot roost on the window ledges or lay their eggs on top of them.

THE EGG STORE

The egg store should be equipped with a cooler to reduce the temperature of the warm eggs to one at which they will not evaporate and lose weight, ideally between 50–54°F (10–12°C). Any loss in average egg weight will cost the producer many hundreds of

An egg room cooler.

pounds over a 52-week laying period (*see* pages 69–70). Retail trade
will also be lost as customers realize the eggs are not as fresh as
they should be. If warm eggs are sent to the packing station the
average egg weight will be lower than it should be and the amount
of second-quality eggs will be higher than normal. It is important
to insulate the egg room well as this will reduce the running time
of the cooler and substantially reduce the cost of electricity.

ELECTRIC FENCING

An electric fence will not only keep the birds safe outside the house,
it can also help to reduce the number of floor eggs that might be
laid inside the house. One strand of 1in (25mm) wide electric horse
tape (white is best) positioned 6in (15cm) off the floor and ½in
(15mm) from the wall will act as a real deterrent to any prospective
floor layer. A good earth is essential, so clip a 10mm wide strip of
chicken-wire to the slatted floor, running directly underneath the
live horse tape, and fix it to a deep earthing pole outside the house.
After peak egg production is reached, the fence can be switched off
and removed, provided there are only a few floor eggs.

The same type of horse tape can also be attached to the outside
perimeter fences, acting as a strong visual deterrent to straying

A non-electric fence will protect the birds from foxes and be safer for any playing pets and children.

birds as they will remember its effect on them from inside the house. Other types of outside fencing available include several different makes of electric netting, and 5ft (1.5m) high wire or plastic fencing. If plastic or wire fencing is used, always put a single strand of electric wire 6in (15cm) above the earth and 6in (15cm) from the fence (on the outside) to stop foxes trying to dig their way in. Keep the fence away from any hedges or trees which could help a predator get over the top of the fence.

FITTINGS AND FIXTURES

Much has been learned in recent years about the importance of getting the lighting right on free range and barn systems. Cannibalism, floor eggs, perching and bird movements at catching times can all be controlled by dimmers and sequential lighting patterns.

Dimmers

A dimmer switch will control any minor cannibalistic tendency before it develops into a serious habitual problem. A gradual reduction in the light intensity by 20 per cent over seven days will

33

also reduce early signs of vent- or feather-pecking. Never increase or decrease the brightness of the lights suddenly, except when catching. A sudden decrease in light intensity will reduce egg production, whilst a sharp increase can trigger feather-pulling, especially at the base of the tail where the feathers are often covered in preening oil. At depletion time the dimmer will play an important part in rounding up the birds ready for catching (*see* page 87). The dimmer switch should be situated on the main control panel and securely fixed to avoid any accidental change in the brightness of the lights.

Sequential Lights

These are lights that are sometimes used to help the birds perch on the A- or H-frames or, in the case of deep litter, on the roosting pit. After the main lights go out, the sequential lights, which are above the desired perching places, will stay on for another 15 minutes. This lighting system helps to stop birds roosting in the nest boxes at night, so reducing the number of dirty eggs. One light per 1,000 birds will be sufficient.

Passageway Lights

These lights should be spaced at 5ft (1.5m) intervals and should be suspended below the tops of the nest boxes, clear of the egg collector's head. If the distance between the lights is increased, shadows cast by the egg collector make it difficult to sort the eggs as they are collected. The passageway lights must have a manual switch so that the collector can use them only at egg collecting times. Because of the possibility of lights being left on by mistake, they should be linked in with the main house lighting. This will guarantee that the controlled day length of the birds is never extended accidentally, causing a complete change in the egg laying times of the birds, i.e. laying all day instead of mainly in the morning.

Main House Lights

Divide the roof of the house into 10ft (3m) 'squares' and place the lights so that there is one in the centre of each 'square'. They should, like all illuminations in any intensively stocked house, be placed as high as possible above the birds. All lights in the house must be shielded to prevent any bright light shining directly on to the nest boxes and flight rails, as layers will naturally choose to lay their eggs in a nest box that is darker than the rest of the house. It is important to make sure the flight rails are included in the darker area as this will assist in reducing any incidence of vent-pecking.

The intensity of light on the nest boxes and flight rails should be between two and five lux, whilst that on the feeding and perching areas should be between five and ten lux. Check the luminosity levels with a light meter before the birds are put in the house. When illuminating a deep litter house, be careful not to create dark shadows along the sides of the roosting pit as this will make the floor an ideal nesting place for the birds and increase the risk of floor eggs. In houses where the stocking density is over 15 birds per sq. metre (12.5 birds per sq. yard) it might be necessary to illuminate the centre of the perching frames, protecting the lights with heavy glass covers to stop the birds pecking and breaking the bulbs. The extra lights will help to illuminate the floor beneath the perches, reducing the risk of floor eggs.

When the poultry house is left unattended a red light outside the house will tell you, at a glance, whether the mains power is on or off. It is essential that there is an alarm system that will alert somebody immediately if there is a loss of power to the poultry house. One very good system is connected to the telephone: in an emergency, pre-programmed numbers are automatically rung in

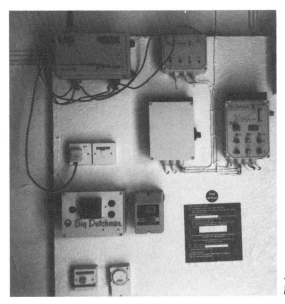

A neat and tidy control panel.

turn until an answer is obtained. A similar telephone system will tell whether the power, heating and ventilation systems are all working satisfactorily when a special number is rung. All these emergency systems and, of course, the generator, require electrical points near to the main control panel. All the switches and controls should be situated in a dustproof box near the entrance to the house.

Electric Lamps

A good choice of very economical lamps is available for use in the poultry house. The most recent fluorescent lamps will reduce the cost of electricity by approximately half, as they only use 20 per cent of the power of conventional lamps to emit the same amount of light. Although they cost more initially, they have a very long life which, combined with the savings on running costs, makes them very cost-effective. Choose one of the warmer coloured lamps as these cast very soft shadows which do not alarm the birds as much as shadows cast by the 'colder' white lights. Do not use clear lamps as the new layers will tend to lay in the darker shadows they cast on the floor, especially underneath the bell drinkers. Make sure all lamps can be controlled by the dimmer.

Feeder Point

The feeder hopper should always be placed outside the laying house, away from the birds. If the feeder is inside the pen with the birds it is not only very difficult to work without the birds getting in the way if any repairs are required, but they will almost certainly lay many eggs in or around the feeder, mainly on the floor.

4

Ventilation Equipment

AUTOMATIC FAN-CONTROLLED VENTILATION

Poultry houses with manure collecting systems are only cleaned out once at the end of the flock (that is, at the end of their useful life, usually 72 weeks), and must therefore have a ventilation system that keeps the air inside the house free from dust and ammonia 24 hours a day. Ammonia levels as low as 10ppm can adversely affect egg size, egg quality and the health of both the birds and the people working on the farm. Any diseases, especially respiratory diseases, will be aggravated by the presence of ammonia in the air. Most of the ammonia problems will occur at night when the temperature drops and the fans go off. Many thermostats are slow to respond to a rise in house temperature so the ammonia builds up to an offensive and sometimes dangerous level long before the fans cut in.

In all deep pit or roosting pit housing flies will be a major problem, breeding in the manure if the air is not flowing continuously over the manure collecting pits. Flies do not like moving air and many of the fly problems occurring on deep pit systems are caused by inadequate ventilation.

It is not difficult to imagine just how unpleasant it must be for the birds sitting above a manure pit at night with little or no ventilation to remove the heat and ammonia. On farms with poor ventilation the loss of income from disease, poor egg quality and low egg production can be very high. Watery whites in eggs is a common problem sometimes caused by a warm stuffy atmosphere in the poultry house. Low house temperatures, on the other hand, will cause an unnecessarily high feed consumption in the birds as they eat more to keep warm. Egg collectors will soon become discontented with their job if they are expected to work in houses smelling of ammonia and infested with flies or red mite, all of which thrive in damp, warm housing conditions. It is therefore important to get the ventilation right, not only for the sake of the birds but also for the farm staff.

The best results are obtained from systems of ventilation which keep all the fans running all the time. All fans will automatically run from full speed in warm weather down to 5 or 10 per cent of their maximum speed when it is cold. This system of ventilation needs no adjustment once the single thermostat is set to the required house temperature. This system of continuous ventilation removes any danger of an ammonia build-up at night.

One 28in (720mm) high speed fan is required for every 1,000 birds housed. These fans must be spaced equally along the top ridge of the roof, avoiding the extreme ends of the house. It is very important that the fans are properly installed, so ask the advice of the makers before starting. All the fans should be linked to one thermostat so that they all operate together, giving equal amounts of clean air to every bird in the house. Each fan should have an independent on-off control and back draught shutters in case it needs to be isolated for repairs. In extremely cold weather it is possible to isolate one or two fans, but they must have back shutters on them to stop the working fans pulling cold air back through the blades of the isolated fan and unbalancing the air movement in the house.

Use a good quality thermostat and include a provision for manual speed control of the fans in case the thermostat breaks down. Place some maximum-minimum thermometers in the house and record changes in temperature daily, taking note of any sudden changes that occur. Adjustments can then be made to the ventilation controls to correct any wide fluctuations. Daytime temperatures in the free range house during the winter will inevitably suffer from the opening of the pop holes, but the night time temperatures should be steady. The barn egg system, using automatic continuous ventilation, should have a fairly constant temperature 24 hours a day.

The ideal temperature for a barn or free range unit is between 64–68°F (18–20°C). Wide variations in temperature can trigger vent-pecking. When the temperature in a house falls suddenly for a period of two or three days, the birds respond by eating more feed. The large increase in feed consumption will suddenly increase the egg size, causing the vents to tear in a few birds. The bleeding vent soon becomes a target for other birds to peck. An increase in blood-stained egg shells can often be related to a sudden drop in temperature four or five days earlier in the week. It is, therefore, vital to try to avoid wide fluctuations in the house temperature.

A naturally ventilated house with sun veranda or scratching area.

NATURAL VENTILATION

The financial advantages of naturally ventilated poultry houses over fan ventilated systems would seem to be obvious: there is no expensive equipment to buy, no running or maintenance costs, and no noise from high speed fans to annoy the neighbours. In contrast to the low cost of setting up the naturally ventilated house, the cost of setting up a fully automatic fan-controlled ventilation system will make a large hole in that already over-stretched bank loan. So why do the majority of egg producers prefer to use a fan-controlled ventilation system?

Although there would appear to be some obvious financial advantages in using natural ventilation, they are not always cost-effective in the long term. Layers in naturally ventilated houses will perform very well but, because of the reduced rate of air changes, the house must not be overstocked. Naturally ventilated houses are better suited to the systems of egg production which have lower stocking densities, such as deep litter or litter and roosting pits. Stocking densities greater than 12 birds per sq. metre (10 birds per sq. yard) are not advisable when using naturally ventilated housing, especially on barn egg systems where there are no pop holes to increase the air movement.

If large eggs are required, i.e. 65g average egg weight at 30 weeks, birds must have a good appetite, and to develop a good appetite the birds must be stimulated by comfortable or cool house temperatures. In naturally ventilated houses this is easily achieved in the winter months when the outside air is cooler than the air inside the house, but it can be impossible to reduce the house temperature during the summer months.

The house design on page 79 can be used in a natural ventilation system. The fans will need to be replaced with a 2ft (60cm) ridge running the length of the house; the ridge vent must be fitted with an adjustable flap which can be fitted to an automatic (hydraulic) or manual control lever; and a 1ft (30cm) deep adjustable window should run down both sides of the house. It is very important that the adjustable windows and the ridge vent are baffled properly to prevent any light creeping in. This is particularly important during the long days of the summer months when the day length inside the house is reduced to nine hours and the natural daylight outside is seventeen hours. Good insulation is necessary in all naturally ventilated poultry houses to achieve a healthy environment for the birds inside the house.

— 5 —

Stocking Densities and Equipment

The number of birds which can be successfully kept in a relatively small area will be determined by management and husbandry skills. If the management is wrong then the higher the stocking density in the house, the greater the problems will be. The effect on bird welfare of increasing the stocking density in a barn or free range house is not completely understood. The following advice, based on fifteen years of practical experience, will, however, help the new or prospective free range or barn egg producer to avoid some of the major pitfalls.

There is no doubt that a bird's performance becomes less predictable if the space in the house in which it lives is reduced. If the number of birds in a house is increased, the amount of stress each bird has to live with is also increased. Density alone is not the only indicator of a bird's welfare; equipment, housing and ventilation all contribute to it. Feeding, drinking, perching, ventilation and nesting equipment must be sufficient to provide all the birds with an equal opportunity to live contentedly and in harmony with one another. An insufficient or unequal distribution of drinking and feeding equipment in a house can cause social instability within a flock of birds, which can sometimes lead to aggression as the dominant birds try to keep the subordinate ones away from the drinkers and feeders. It is, therefore, important to remember that the equipment and the evenness of its distribution in the house become more critical as the stocking density is increased. The following guidelines will give all the birds in the house, no matter what part of the house they occupy or what the stocking density is, a fair share of the feed and water.

DEEP LITTER

7 Birds per Sq. Metre (5.8 Birds per Sq. Yard)

Partition the birds off in pens of 1,000 to 2,000 birds, as this will stop them drifting to one end of the house. Feed and water must be available on both the roosting pit and the deep litter area. It is convenient to put all the feeding equipment on the roosting pit and the drinking equipment on the litter area or vice versa, but if you use this system make sure all the birds are finding both feed and water. This is especially important when new birds are put into a house at 18 weeks. All the birds must be moved off the roosting pit every morning, and any birds on the litter 30 minutes before the lights go out at night must be put up on to the roosting pit. The higher the stocking density the more important it is that this

A deep litter system with a roosting pit.

arduous task is carried out, as some subordinate birds will prefer not to drink or eat and will become weak or even die.

The ideal system of deep litter egg production will have 33 per cent roosting pit space and 67 per cent deep litter area. Larger roosting pits, using up more than 60 per cent of the floor area, are inclined to encourage floor egg problems, especially if the nest boxes are positioned in the litter area. It is better to have an all slatted floor than a very large roosting pit.

PERCHERY OR BARN

Up to 15 Birds per Sq. Metre (12.5 Birds per Sq. Yard)

In the perchery or barn egg system, with a stocking density of 15 birds per sq. metre (12.5 birds per sq. yard), the feed and water can all be on one level, on the floor. The correct number of bell drinkers must be used on the floor (*see* page 50). Extra nipples situated on the underside of the higher A- or H-frames will help to satisfy the birds' thirst in the warm summer months when water consumption can double. Feeding and drinking equipment must be spread out equally in every part of the house so each bird has the same distance to walk to get feed and water.

15 to 25 Birds per Sq. Metre (12.5 to 20.9 Birds per Sq. Yard)

In these high stocking situations the easy availability of feed and water to every bird in the house is critical if good results are to be obtained. Even so, experience has shown that with houses stocked at over 20 birds per sq. metre (16.7 birds per sq. yard) the results are most unpredictable. Feed and water must be provided on at least two or three levels. The best results come from using bell drinkers wherever possible, although nipple lines attached to the underside of the perch racks work well. There is no doubt that poultry prefer open-sided drinkers (bell type) when given a choice. Nipple drinkers can be used with cups for the higher levels. The advantage of the cups is that they can be filled with water, which gives the birds somewhere to drink while they get used to using the nipple drinkers as the majority of 18-week-old birds will not have used a nipple before.

Failure to make sure that feeding and drinking equipment is sufficient and easily available to all the birds can reduce egg size and encourage aggressive behaviour in a flock of birds housed at this density.

FEEDING EQUIPMENT

Automatic Flat Chain Feeder

The automatic flat chain feeder is the most popular feeder in use today. It is the cheapest and easiest to install and, provided it is set up properly, is reliable and needs very little maintenance. It is controlled by a time clock and should be set to run from 5 times per day at 18 weeks of age to 8 times per day at 25 weeks. Feeding space should be provided at 4in (10cm) per bird, i.e. 6 birds per 1ft (30cm) of feeding space, allowing for the fact that the birds

The higher level of feed track required in houses that have a stocking density in excess of 15 birds per square metre.

The special rising bracket required to take the feed up to a higher level.

can feed from both sides. The feeder should be positioned so that it is at the same level as the birds' backs. If it is any lower the birds will waste the feed by flicking it out of the trough. Low feeders will also cause an obstruction to the birds' route to the nest boxes which encourages them to lay on the floor. The feeder can be connected to a winch system to lift it clear at cleaning-out time. Do not put too much feed on the chain as this allows the birds to dig it out and waste it. If the feeder keeps returning the feed and there is a build-up as it goes back into the box, the outward levels are set too high.

The automatic flat chain feeder is ideal for stocking densities over 15 birds per sq. metre (12.5 birds per sq. yard) as it can be used at many different levels. One of the disadvantages of this feeder is that it must be repaired very rapidly if it breaks down as it only holds enough food to feed the birds for two or three hours.

Automatic Overhead Feeders, Filling Tubes or Pans

Second in popularity is the overhead tube feeder, which has some advantages over the flat chain feeder. The overhead feeder allows

A centreless auger delivering feed overhead to tube feeders in the house.

a lot more time for repair as it holds much more feed in reserve; large tube feeders have reserves of feed to last up to 24 hours, although smaller pan feeders do not last as long. Another advantage is that there is no need to continually check for feed in the tube feeders or pans. The disadvantages of the overhead feeder are the initial cost of the equipment, which is approximately 33 per cent higher than the flat chain feeder, and that it cannot be used for higher levels in a highly stocked perchery or barn system.

Once the overhead auger is fixed permanently in place one tube or pan feeder per 25 birds will be required. The height of the feeders should be set so that the top feeding lip of the pan is level with the birds' backs. If the pan feeders are too low not only will feed be wasted, but a dark shadow will be cast making a nice nesting place for floor egg layers. As with all feeders, keep the feed level low to prevent wastage by the birds flicking the feed out.

Keep the feeder away from the birds by concealing it. This house has A-frames and plastic floors.

Manually Fed Tube Feeders

Like all tube or pan feeders, these can be made of plastic or metal. They are very good for the small commercial egg producer, as 2,000 or more layers are required to justify the cost of installing an automatic feeder. The modern plastic feeders have the feed levels pre-set ready for use and are made in a variety of colours. Four manually fed tube feeders per 100 birds is normally adequate. As with all feeders, the top feeding lip should be level with the birds' backs.

Bulk Bins

Most bulk bins are made of metal, although some are made of glass-fibre. The glass-fibre bins are more expensive but have the advantage over metal that it is possible to see how much feed remains in the bin.

To transfer feed an auger can be used, or the bin can have a long side discharge going straight into the feeding system inside the house. The advantage a gravity-fed side discharge bin has over an auger discharge bin is that there are no expensive mechanical parts to go wrong. Augers are expensive to install and run, but have their advantages, especially if two houses are run from one bulk

47

A gravity-fed side discharge bulk bin.

bin. When buying a bulk bin make sure it will hold enough feed to last the birds for at least two weeks; not only will this cover holiday periods and so on, but if you buy larger loads of feed you should qualify for a price discount from the supplier. Ensure the bin is designed for meal rather than pellets.

Drinkers

There are two types of drinker used on commercial egg farms: the nipple and the bell drinker. Given a choice, most animals or birds prefer to drink from an open drinker, because of their natural habit of drinking from ponds, rivers and puddles.

For deep litter systems the bell drinker is best, not only because the birds prefer them but because on an all litter or litter and slatted system there is no need to use nipples as the drinkers will all be hanging at floor level. Nipples are best used for higher levels on A- or H-frames where it is difficult to use bell drinkers. Bell drinkers, like all feeding and drinking equipment, should be raised off the floor to the height of the birds' backs. The water level should

Side feed discharge bin with bagging-off facility.

Tube feeders. The larger type are used for automatic overhead feeding systems.

49

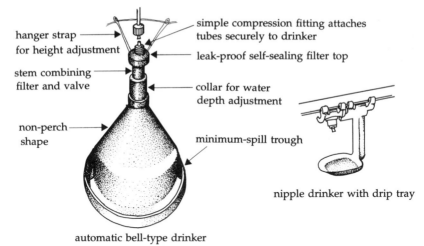

Bell drinker and nipple drinkers.

be set half-way up the drinking rim of the bowls. If there is too much water the birds will flick and dribble water everywhere, making the litter (or manure in the deep pit) wet and increasing the levels of ammonia in the house. Wet manure will also make cleaning-out difficult.

One bell drinker for every 100 birds is normally adequate, as is one nipple drinker for every five birds. When using nipple drinkers make sure that the birds can reach them easily. Provided the vertical distance between perches is no more than 16in (40cm), the birds will be able to reach the nipples easily. Always use nipple drinkers with a cup underneath.

Guide to Water Consumption for 1,000 Birds

Age in Weeks	Litres	Gallons
4	50	11
8	90	20
12	115	25
16	140	31
20	170	37
24	195	43
30 to end of lay	205 to 225	45 to 50

These are only approximate water intakes as they will vary according to house temperatures in the barn system and the outside temperature if the birds are free range. In winter the birds will drink less and in the summer they will drink more.

PERCHING FRAMES

Perches are essential in all free range or barn egg producing systems with a stocking density in excess of 7 birds per sq. metre (5.8 birds per sq. yard), including the roosting pit area in any deep litter system (*see* perching regulations, pages 106–107). Perches get birds up off the floor, reducing the competition for feeding, drinking and floor space, which is an important factor in relieving the stress and aggression that can develop in highly stocked egg producing systems.

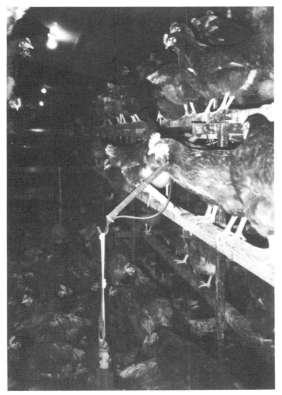

An H-frame.

— 6 —

Rearing

CONTROLLED ENVIRONMENT REARING HOUSES

The way in which an 18-week-old point of lay (p.o.l.) pullet has been reared can determine the degree of success or failure it will have during its 52-week laying period. It is important, therefore, when buying birds from a rearer, to make sure that they have been reared in a controlled environment house with full control of the lighting, feeding and temperatures.

Cage or Deep Litter Reared

Birds which are going to be used for free range, barn or perchery egg production must be reared on deep litter, not in cages. Birds reared in warm, movement-restricting cages cannot develop the good appetite which is necessary to sustain good egg size and high production. Also, the free range or barn egg layer needs to develop a high degree of immunity from the diseases and parasites it might encounter in the field, such as coccidiosis and worms, and birds reared in cages cannot develop any immunity from these.

Deep litter reared birds will be more familiar with deep litter or free range laying houses and will soon settle down, but cage reared birds are very slow to adapt to their new surroundings. Cage birds will have little or no idea how to perch and they do not like to use the high nest boxes or high perches. This can add to the problem of floor eggs as the birds find it difficult to find a place in the over-used bottom nest boxes and so lay on the floor.

How Light Controls Egg Production

Egg production is closely related to the changes in day length to which the chicks (day-old to 6 weeks), growers (6 to 16 weeks) and pullets or layers (16 weeks onwards) are exposed. Egg numbers,

Point of lay pullets quickly find the perches after housing.

egg size and eventually the amount of profit or loss achieved from layers can be determined by the lighting pattern used in the rearing stage. Wild birds experience a naturally decreasing day length between the summer and winter solstices, which will eventually stop the bird laying. This is nature's way of giving the mother hen a rest, during which time she will use up the excess egg producing energy by renewing her feathers (known as the moult). She will then be ready to start laying eggs again and, as the daylight hours increase, will be stimulated into lay. This is the reason why all wild birds start to lay in spring. The important lesson to be learned from this is that decreasing the light to a bird will reduce or stop the production of eggs and increasing the light will start or increase the production of eggs. Never, therefore, reduce the hours of light to a bird in lay.

The rearer uses artificial light in a similar way to control the maturity of the birds being reared. By gradually reducing the day length from 23 hours at day-old down to 8 hours at 8 weeks of age,

53

the birds are given plenty of time to eat and fully develop but their maturity (laying development) is delayed, preventing them from coming into lay too early. This lighting pattern will produce plenty of large eggs provided the birds are given a proper lighting programme in lay similar to the Grassington Rangers' programme shown in the table below. There are other rearing lighting programmes which have been designed to reduce egg size by bringing the birds into lay a little earlier with a reduced body-weight. So there is a rearing lighting programme to help achieve the egg size

GUIDE TO ARTIFICIAL DAYLIGHT LENGTH														
Month	Natural day lengths	Wk 18	19	20	21	22	23	24	25	26	27	28	29	30
Jan	8–9	9	10	11	12	13	13	14	14	15	16	——→		
Feb	9–10.5	9	10	11	12	13	14	14	15	15	16	——→		
Mar	11–13	9	10	11	12	13	14	15	15	15	16	——→		
Apr	13–14.5	9	10	11	12	13	14	15	16	——→				
May	15–16	10	12	14	14	15	15	15	16	——→				
June	16–16.5	10	12	14	15	15	15	16	——→					
July	16.5–15.5	10	12	14	15	15	15	16	16	——→				
Aug	15–14	10	12	14	15	15	15	16	16	——→				
Sept	13.5–11.5	10	12	13	14	15	15	15	15	16	——→			
Oct	11–10	9	10	11	12	13	14	15	15	15	16	——→		
Nov	9.5–8.5	9	10	11	12	13	14	14	15	15	16	——→		
Dec	8–9	9	10	11	12	13	13	14	14	15	16	——→		
(For birds housed at 17 weeks)														

Grassington Rangers' lighting programme, which is designed to make sure the birds receive adequate light stimulation at all times of the year, in particular the autumn months when natural daylight is reducing.

0–3 days	22 hours	
3–6 days	20 hours	
7–21 days	18 hours	
22–35 days	16 hours	
36–49 days	14 hours	rearing period
50–56 days	12 hours	
57–70 days	10 hours	
10–17 weeks	8 hours	
18 weeks	9 hours	
19 weeks	10 hours	
20 weeks	11 hours	
21 weeks	12 hours	laying period
22–23 weeks	13 hours	
24–25 weeks	14 hours	
26 weeks	15 hours	
27 weeks	16 hours	

This rearing and laying lighting programme is for January p.o.l. birds only, and is designed to delay maturity in order to produce larger eggs.

required, and chick breeders or poultry specialists will advise on this.

With complete control of the young birds' environment, the rearer is able to limit undesirable influences from the outside, such as stray shafts of light which will stimulate the birds into laying early. The amount of luminosity needed to stimulate a bird into lay is only 0.4 lux, the brightness of moonlight. So the influence of light on young birds and their future egg size is very important.

Body-Weight

A controlled environment rearing house will give the rearer complete control over the body-weight of the young birds. The more uniform the birds are at 18 weeks of age the better they will perform

1–2 days	22 hours	
3–4 days	20 hours	
5–6 days	18 hours	
7–8 days	16 hours	
9–10 days	14 hours	
10–12 days	12 hours	Rearing period
13–14 days	10 hours	
2–17 weeks	8 hours	
18 weeks	9 hours	
19 weeks	10 hours	
20 weeks	11 hours	
21 weeks	12 hours	
22 weeks	12.5 hours	
23 weeks	13 hours	Laying period
24 weeks	13.5 hours	
25 weeks	14 hours	
26 weeks	14.5 hours	
27 weeks	15 hours	

The ISA Brown lighting programme, which should only be used in perchery or barn systems. This lighting programme is geared to producing optimum egg numbers on target egg weight for age. For larger eggs or for free range egg production, use the lighting programmes on pages 54–5.

in terms of peak egg production and laying persistency. Flocks which have not grown to correct body-weight and have unacceptable levels of uniformity will not perform to their full potential and will exhibit the following production trends:

1. Production increases at the onset of lay will be slow.
2. The interval between first egg and peak production will be extended.
3. Low flat peaks will occur.
4. A wide variation in egg size will develop.
5. Feed consumption will be depressed because of the smaller birds.
6. The persistency of egg production will be reduced.

The ideal 18-week-old flock of birds will be judged more by its evenness than by average weight. At 18 weeks flocks should achieve a uniformity of 85 per cent, plus or minus 10 per cent of the average flock weight. Very small birds should be rejected. Account should be taken of the weight loss suffered because of the loading, unloading and travelling between the rearing and laying farms. It normally takes seven to ten days before the birds' body-weight begins to increase again, owing to the stress of being moved and settling down in the new house.

Feeding Programmes

During the rearing period the birds will be fed two or three different types of feed. The standard feeding programme is usually chick starter feed (mash or crumbs) to six weeks of age, followed by grower mash to eighteen weeks. An alternative is to use a pre-layer ration at sixteen weeks. This is designed for feeding immediately prior to lay and includes energy, protein and amino acid levels at higher levels than the grower feed to help the birds to maintain high peak production, plus good egg size. The birds should be fed ad lib throughout the rearing period.

All the feeds (except sometimes the pre-lay) contain an anti-coccidiosis drug to help the birds to control and develop their own immunity to coccidiosis.

Vaccination

The birds should already be fully vaccinated when you receive them. The normal vaccination programme provides good protection against infectious bronchitis (IB), Newcastle disease (ND), egg drop syndrome (EDS) and epidemic tremors (ET). There are vaccines available for other diseases but they are only used when it is known that those particular viruses or bacteria are causing problems in or near the area of destination. A specialist poultry vet will advise what extra vaccinations might be required.

There are two different types of vaccine in general use. The first is a live vaccine containing a live virus of a specific disease which is weaker than the real field challenge which can affect your birds. After administration, usually through the drinking water or spray, these live organisms will stimulate antibodies in the birds providing them with good levels of immunity against a real field challenge. The second is a dead virus mixed with a special oil and is administered by injection into a muscle in the leg or breast. These vaccines

are called inactivated vaccines. The combination of using live vaccinations to prime the birds in rearing followed by a dead vaccine to
boost immunization at 18 weeks (p.o.l.) will give birds good protection against IB and ND. Rearers will normally provide a list of all
the vaccinations given to the birds and some will provide the results
of blood tests which show the levels of immunity the birds have
developed through the vaccinations they have received. If you are
rearing your own birds, ask a vet for a full vaccination programme
well before the chicks arrive.

Perches

There is no doubt about the many benefits to the commercial egg
producer of birds which have been perch trained in the rearing
period. Birds that have not been reared with perches in the house
are more likely to stay at floor level, causing overcrowding and
intense stressful competition at the feeders and drinkers; sit on the
feeders or anything else that is not too high off the ground; sleep
or perch in the lower nest boxes if left open at night, fouling the

*Perch training is an essential
part of the rearing
programme.*

nest and causing more dirty eggs; and lay in the bottom nest boxes causing severe overcrowding at peak laying time. It will also be difficult to walk around the floor of the house whilst looking for floor eggs and removing any dead birds.

REARING YOUR OWN BIRDS

Most egg producers buy their birds from an established rearer, but a few prefer to rear their own and, provided they have a good rearing house, this does have some advantages. If the birds are reared in the same area as their laying house they will get used to all the local viruses before they come into lay, although they will still need a full vaccination programme for which you should consult your vet. Also, the birds will not have far to travel when they are housed, which means low transfer costs and less stress to the birds. The two main disadvantages are the setting-up costs and the probable lack of experience of the rearer trying to do the job of a specialist.

The Rearing House

Stocking Density, Feeding and Drinking Space and Litter

The rearing house must be well insulated, well ventilated and lightproof. Allow the birds 0.85ft² (790cm²) each in a fan ventilated house (one 28in (720mm) fan per 1,000 birds). In a naturally ventilated house, stock at 1ft² (929cm²) per bird. Whatever the ventilation 3in (7.6cm) of feeding space will be required if using a flat chain automatic feeder, i.e. 8 birds per ft (30cm). If tube feeders are used, one plastic tube (pre-set for feed depth) per 25 birds will be required. One hundred birds will require one plastic bell-type drinker, which should have a shallow drinking bowl for the day-old chicks to drink from. Ensure that the water tank supplying the automatic bell drinkers is large enough to hold at least 12 hours' water for the birds in case of a mains water failure or a need to medicate the birds' drinking water. The house and equipment must be clean, and white softwood shavings should be spread over the floor at a depth of 4in (10cm) and inside the brooder surround at a depth of 6in (15cm).

Brooders and Brooder Surrounds

Gas brooders emit a warm moist heat which promotes good feather-

Hand founts for chicks and small-scale egg production.

A

A easy-to-fit clamp fixing for water pipe

B automatic bell shape bowl drinker

C small chick drinkers for the first 3–4 days

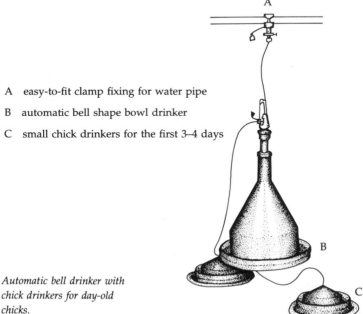

B

C

Automatic bell drinker with chick drinkers for day-old chicks.

A gas brooder.

ing, and continue to operate during a power failure. Electric brood-ers emit a dry heat, which can make the feathers brittle, and will go cold as soon as the electricity goes off.

Place the brooder in the centre of the house, away from cold walls or windows. Use corrugated paper or hardboard strips 2ft (60cm) high to make a circular surround 10ft (3m) in diameter with the brooder hanging in the centre of the surround. This will keep approximately 2,000 chicks around the brooder for four to five days, depending on the litter condition and outside temperatures. The surround must be enlarged as the birds get bigger and removed by one to three weeks of age, depending on the temperatures on the other side of the surround.

Preparing the Brooding Area
Prepare the equipment inside the surround 24 hours before the chicks arrive, checking that the brooder is working satisfactorily. Small light bell drinkers can be used to give the chicks their water on the first day. When setting up the house, put in some extended drinker supply pipes which can be used in the surround during the first week and then moved out with the chicks as they spread across the house. Drinkers must be placed as near to the floor as possible so that the chicks can reach them easily. The water level

in the bowl should be high so that the chicks can see the reflected light in the water, which will encourage them to drink.

When the chicks are first put under the brooder it is better that they drink first and then eat, especially if they arrive on the farm tired after a long warm journey. If tired, thirsty chicks eat first, the food will act like blotting-paper causing dehydration, sometimes leading to 'non-starters' or death. So, for the first two hours allow the chicks access to water only and then distribute the feed. Clean egg trays can be used to hold the feed – 15 egg trays per 1,000 birds – and these should be placed firmly in the litter so that the birds cannot get underneath them. Food and water must be spread evenly all over the brooding area to enable the chicks to eat and drink anywhere in the surround. Adding multi-vitamins to the water for the first 48 hours helps to reduce early chick mortality, especially if the chicks have been stressed in transit. Always make sure the drinking water has been warmed up overnight before the chicks drink it.

Set the brooder thermostat so that the temperature under the brooder is 95°F (35°C) and 90.5°F (32.5°C) at the edge of the surround. Approximate brooder and house temperatures are given in the chart below, but these are only a guide; bird behaviour and movement will help to confirm whether the temperature is correct.

APPROXIMATE BROODER AND HOUSE TEMPERATURES AND LIGHT INTENSITY IN REARING

AGE AND TEMPERATURE		INTENSITY IN LUX
1 day	95°F – 35°C	20 – 40
1 week	91.4°F – 33°C	20 – 30
2 weeks	86°F – 30°C	10 – 20
3 weeks	82.4°F – 28°C	5 – 10
4 weeks	77°F – 25°C	5 – 10
5 weeks	73.4°F – 23°C	5 – 10
6 weeks	70°F – 21°C	5 – 10
7–18 weeks	66°F – 9°C	5 – 10

The Chicks' Arrival

Twenty-Four Hours to Three Days
When the chicks arrive take them straight from the van and empty

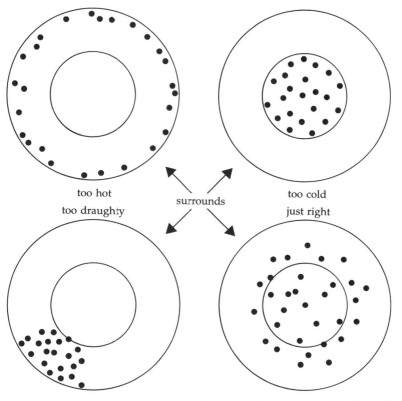

too hot
too draughty

surrounds

too cold
just right

chick movement is your best guide to how comfortable the chicks are

Bird behaviour under the brooders.

them out carefully near the centre of the brooder. An hour spent with them making sure that none of the drinkers are overflowing or are too high for the chicks to reach is a good way of getting them used to you. After making sure the brooders and drinkers are working satisfactorily turn the lights off for a minute to accustom the chicks to the darkness. This will prepare the birds for the increasingly long dark periods they will experience as they are introduced to the lighting programme using the automatic time clocks.

Continually check the birds' behaviour to make sure they are not too hot or too cold. If the chicks show distinctive signs of being too cold, such as huddling together, turn the brooder heat up. If they are all moving away from the centre of the brooder and show signs of panting, slowly turn the heat of the brooder down (see page 63).

Check feed and water levels throughout the first day and for the first three days, replacing feed and water as and when required. Clean drinkers daily using a bucket to keep the litter dry.

Three to Nine Days

At this stage begin to replace the egg trays with some tube feeders at the rate of two feeders per 1,000 chicks each day. If an automatic flat chain feeder is being used, lower the feed trough down on to the litter and send some feed round, switching the feeder on by hand. This will get the birds used to the noise. Some rearers prefer to lower the feeder on to the litter at day-old, even though the chicks do not use it much, to give them an early opportunity to become accustomed to it. When the feeder has been lowered, place half of the egg trays alongside the feeder track to encourage the birds over to the feeder to feed. By the fifth day all the egg trays should be moved near the feeder track. At six days the chicks must be beak trimmed, after which it is wise to give them some multivitamins in the water. Do not remove any more egg trays until 36 hours after beak trimming. Gradually remove all the egg trays so

A plastic bottle stops the chicks getting caught in the track feeder.

that by the ninth day the chicks have only the track feeder or tube feeders to eat from. Set the feeder time switch to give the birds four feeds per day.

During this period it is necessary to enlarge or remove the surround as the chicks grow and to reduce the temperatures as indicated in the chart on page 62. When to enlarge the surround and alter the temperatures will often depend on the temperatures that can be achieved inside the house. When the birds are released from the surround they must have access to some A-frame type perches.

The drinkers should be cleaned out every day using a bucket to keep the litter dry.

Nine Days to Six Weeks
During this period the feeding times of the automatic feeder can be increased to six feeds per day and the surrounds will have been removed. All the feeders and drinkers should be placed evenly over the house and a walk through the birds two or three times per day to ensure everything is working all right is all that is needed. Clean the drinkers each day using a bucket to keep the litter clean.

Six to Eighteen Weeks
At six weeks (sometimes eight depending on the feed brand used) change the feed from chick starter feed to grower. Before ordering feed weigh the birds and if they are under- or overweight ask the advice of the feed supplier's poultry specialist about adjusting the feed programme. The chick salesman will have a weight-for-age guide for the breed of bird being reared. Try to avoid any other changes in the chicks' daily routine, such as vaccinations, when changing the feed as too much stress can sometimes cause health problems. At 16 weeks change the feed to a pre-lay ration (see Chapter 7). In this last stage it is advisable to constantly weigh the birds so that any adjustment to the feeding programme can be made to correct a weight problem before the birds are put into the laying house.

At 16 weeks take some blood samples and send them to a vet who can analyse the effect of the vaccinations which have been done and make any late amendments if they are required. Taking blood samples is best learned by watching the vet who, in turn, can make sure you are doing it properly. The blood is usually taken from the brachial vein in the wing.

7

Feeding the Commercial Layer

The free range or barn egg producer should always bear in mind that the quality of feed given to birds is very important. For the birds to be successful they must be able to perform to the following standards:

1. Maximum egg numbers, including high peak production and good persistency.
2. Good early egg size with a minimum of small eggs for the large egg market or maximum numbers of medium-sized eggs for the fixed price or catering market.
3. Good internal and external (shell) quality to the end of lay.
4. Food conversion efficiency, producing more from less.
5. Low mortality.

The egg market, and especially the need for either large or small eggs, can influence the feeding regime for free range birds, and to some considerable degree it is possible to adapt the birds' production to the requirements of the market.

SYSTEMS OF FEEDING

For economic egg production and in order to obtain the maximum performance from your birds, ad lib feeding is essential. From the point when the light is increased for the first time, the birds' sexual and physical maturity development will accelerate. Ad lib feeding to the end of lay is essential to peak egg mass and to maintain maximum egg output and weight.

In order for the birds to lay larger eggs they require a good appetite. It is possible to stimulate the birds' appetite by feeding a low energy feed, because the energy content of feed will dictate to a large extent how much a bird will eat: the lower the energy of the feed, the more a bird will eat. The low-energy feed will reduce the risk of the birds putting on large amounts of fat when they are only producing a few small eggs. This will help to prevent the birds from prolapsing. In contrast, high-energy feeds cause a build-up of fat around the pelvic area which makes the passage of the egg past the pelvic bones very difficult, and often leads to prolapsing or cannibalism. Low energy feed should only be used from approximately 18 to 21 weeks. At 21 weeks the birds should have developed a good daily feed intake and a change to a high energy, good quality feed will give the birds all the nutrients they require to sustain a high peak and good egg size. A high energy, high specification ration can be fed to the end of lay. If, near the end of the laying year (after 60 weeks), the eggs are becoming too large and the shell quality poor, change to an end of lay ration, which is specially formulated to improve shell quality in old birds.

Another successful feeding programme for large eggs is to start the birds at 18 weeks on the high energy, high specification food and feed it throughout lay. Use a mid-energy ration (11.3ME) throughout lay for smaller eggs. One other popular feeding programme is to phase-feed the birds: use an early lay, high specification, high energy ration up to approximately 42 weeks in lay then change to a middle energy, good quality ration, then change again to the end of lay ration at approximately 60 weeks of age.

The most consistently good results come from feeding a good quality, high energy ration throughout lay, using an end of lay ration to help to improve egg shell quality if it is a problem. It is sometimes necessary for the feed to be changed because the egg size is becoming too large, but it is better to make the feed change when it looks like being a problem not because the feed plan says so. There are too many variables to consider in the management of free range egg production, including housing, stocking densities and seasonal temperature changes, without carrying out a pre-arranged timetable of changes to the birds' feeding programme.

THE DIFFERENT FEEDS AVAILABLE

Most feed manufacturers can offer a range of diets to suit all of the free range and barn egg laying systems, ranging from the high density diets designed for the more intensive systems to the lower density diets for the small backyard and farmyard egg producers. Each system requires a different type of feed. The more intensive systems stocked in excess of 15 birds per sq. metre (12.5 birds per sq. yard) require a feed that has a high specification for amino acids, protein, energy and linoleic acid to support very high levels of performance under more stressful conditions. This diet will help to maintain energy intake, peak production and egg size in the smaller birds within these large flocks where competition for feed and water can be fierce. Higher temperatures in the highly stocked houses sometimes reduce the birds' feed intake, making it difficult for them to consume enough nutrients.

In stocking densities of less than 15 birds per sq. metre (12.5 birds per sq. yard) a middle specification feed will be adequate as there is not the same degree of competition for feed and water amongst the birds. With a lower stocking density the temperature in the house should be lower, which will stimulate feed intake. The birds' higher feed intake means that they will get enough nutrients from the middle specification food.

Backyard hens and other small numbers of birds all do well on a low specification feed.

The chart opposite lists a range of the feeds used on free range egg production systems and the different levels of oil, protein, fibre, ash and energy (ME) levels.

Feed Cost

It can be seen from the feed specification guide that the higher the feed specification, the more oil, protein, methionine (Meth.) linoleic acid (Lin.) and metabolized energy (ME) are included. All these ingredients are responsible for improving early egg size, high peak production levels, good persistency in lay and lower daily feed intake. Feed cost is an important consideration in its own right, but more critical are the differences between the varying nutrient specifications of the feeds. Where these differences are small it will be more profitable to feed high specification diets, as the higher cost per tonne will be more than offset by the lower daily feed consumption and improved egg size.

The Cost of Increasing or Decreasing Average Egg Weight
Looking at the average prices being paid by an egg packing station in 1993 and using those prices in conjunction with a well-known commercial breeder's predicted weekly egg grades, we can see how average egg weight can affect a producer's income.

Oil	Protein	Fibre	Ash	ME	Lin.	Meth.	Benefits
High Specification							
6.5%	18.5%	3.75%	13%	12%	2.8%	0.43%	Suitable for all very intensive free range or barn systems which are stocked over 15 birds per m² (12.5 birds per sq. yard)
Mid Specification							
5.5%	18%	4.3%	14%	11.3%	2.1%	0.40%	Suitable for all intensive barn, perchery and free range systems stocked between 7 and 15 birds per m² (5.8 and 12.5 birds per sq. yard)
Low Specification							
4.5%	17%	4.8%	14%	10.8%	1.3%	0.36%	Suitable for all backyard, deep litter, low stocking density systems up to 6 birds per m² (5 birds per sq. yard)

The three main types of feed.

Average Egg Weight	Average Price Per Dozen	Difference
51g	31.5p	
53g	36.8p	5.3p
55g	40.2p	3.4p
56g	42.8p	2.6p
57g	46.9p	4.1p
59g	51.7p	3.8p
60g	54.7p	3.0p
61g	57.5p	2.8p
62g	60.6p	3.1p
63g	63.6p	3.0p
64g	66.4p	2.8p
65g	69.1p	2.7p
65.5g	70.2p	1.1p

Average price difference between grades = 3.14p per per dozen

Cost of loss of egg weight.

If a bird lays 24.5 dozen eggs in fifty-two weeks, the approximate profit or loss of income for each gramme of egg weight would be:

0.5g	38.46p per bird or	£384.60 per 1,000 birds
1g	76.93p per bird or	£769.30 per 1,000 birds
2g	153.86p per bird or	£1,538.60 per 1,000 birds

The Cost of Reducing or Increasing Daily Feed Intake

The lower specification of the feed is not the only reason why the birds' feed consumption may go up; cold housing, parasites and wastage from feeders that allow the birds to flick the feed on the floor will all increase feed usage. Very warm houses, disease and unpalatability will reduce feed consumption.

The following costs would be incurred by increasing daily feed intakes at £170 per tonne:

An extra 1g of feed per bird per day for 364 days = 6.18p per bird.
An extra 5g of feed per bird per day for 364 days = 30.90p per bird.
An extra 10g of feed per bird per day for 364 days = 61.80p per bird.

Palatability

This is a very important factor in deciding how much food a bird will consume with satisfaction. The greater the fibre content of the meal, the less palatable it becomes. Dry dusty mashes are very unpalatable to the birds and will cause a severe reduction in the amount of feed they can eat. If the feed is not removed from the birds, both egg production and size will be reduced.

Physical Forms of Feed

Meal

Meal is any grain in milled or ground form that has not been formed into any size or shape.

Mash

This is a mixture of meals balanced for the different classes of stock and stocking densities. It helps to overcome boredom in a flock as the birds spend a lot more time picking it up and then drinking because it is dry. Mash is the cheapest form of feed available because of its relatively simple manufacturing process.

Crumbs or Peckets
Meal in tiny broken pellet form, crumbs or peckets are usually fed to chicks to improve their feed intake as they are very palatable. They can also be used to increase feed intake if it is low because of hot weather or if the birds are underweight at point of lay. Crumbs or peckets cost more to manufacture and are therefore more expensive than other feeds. However, they can both be fed throughout the life of the flock and the extra cost will be offset by the slightly larger eggs produced. Nevertheless, feeding crumbs throughout lay is not advised as the birds are more susceptible to boredom and cannibalistic problems.

Pellets
These are a balanced meal manufactured into cylindrical pellets varying in size. Pellets are not advised for layers living in intensive situations because of the problems associated with boredom, overweight and cannibalism, but can be fed to small densities of birds, such as backyard hens, or as a scratch feed for the layers on deep litter systems to help to keep the litter moving as the birds scratch to find the feed, i.e. 5g of pellets per bird per day.

Grist
Grist refers to the evenness of the mash. If the grist of the mash is too fine, the birds will eat less feed because it is unpalatable. If the mash is too coarse, selective feeding can take place as some of the birds pick out only the whole grain. The bulk bin should be cleaned out on a regular basis to avoid an accumulation of fine feed as this will depress egg quality and production if the birds eat it for two or three consecutive days. Food with a fine grist will sometimes bind together and clog up the augers and feeders.

Yolk Colourants (Pigmenters)
Yolk colour is measured on the Roche Scale, which graduates from pale yellow (No. 1) to deep orange (No. 14). The colour range preferred by the majority of egg consumers is between 9 and 11 on the Roche Scale. This is a mid-orange colour and is achieved mainly by natural pigmenters called xanthopylls, which are found in maize, maize by-products, grass, lucerne, marigolds and paprika. Artificial xanthopylls called nature identical colourants can also be used.

Calcium
Calcium is added to the feed in the form of limestone granules so, under normal conditions, no extra calcium should be given to the

birds. There are, however, times when shell quality can be strengthened by giving the birds oyster shell grit. Advice on exactly how much can be given can be obtained from the feed company's poultry specialist.

Granite Grit
Hen-size insoluble granite grit is given to the birds to help them grind up their food and can be valuable in helping to prevent compacted gizzards, when the birds have eaten long coarse grass which can block up their gizzards. It is not included in the feed and should always be available to the birds from gravity fed hoppers, although free range birds on stony land will not usually require it.

Meat and Bone Meal
Meat and bone meal has been used as a good source of protein in poultry feeds for many years. Today there is a choice, as some feeds are vegetable protein only. In its wild and natural state the chicken eats both animal and vegetable proteins, so it makes sense to feed them a ration that contains both. Some egg packing stations, however, will not buy eggs that have come from birds fed on animal proteins, so check the contract before ordering feed.

HOME MIXING

The capital cost of setting up your own feed manufacturing mill would deter most prospective commercial egg producers. Nevertheless, there are some advantages in milling your own feed. Apart from all the licensing red tape regarding adding medicines to the feed and so on, the main advantage is that the feed can be made more cheaply because there are no haulage or administration costs. It is also convenient to be able to make the feed as and when it is required. The modern hybrid layer producing eggs in the new highly stocked perchery, barn and free range units requires special high energy feeds that are sprayed with soya oil, and this requires special equipment which the standard home mixing mill does not have.

— 8 —

Systems for Keeping Commercial Laying Hens

DEEP LITTER EGG PRODUCTION

Deep litter egg production is a very successful method of producing eggs, but the day-to-day management of the farm needs to be very good. Any system that includes a deep litter or scratching area inside the house will be much more susceptible to floor eggs than a system without any litter. A bird's natural nesting requirements in the wild, where she would much prefer a cosy grass-lined nest, are just like the litter on the floor, not the plastic-covered floors of the modern roll-away nest box. Deep litter, however, does keep the birds occupied during the day and any scratching or dust-bathing area helps to relieve boredom and aggression in a flock of birds.

Types of Litter

Chopped straw, peat, white softwood shavings, sawdust and paper can be used as litter. All need to be maintained in a dry and friable state in order to 'work' (move freely). Soft white woodshavings give the best results; coloured or hardwood shavings do not 'work' (are less friable) and can colour the eggs so they should not be used. The litter must be at least 4in (10cm) deep from the first day the birds are put in the house. It is no good starting with 2in (5cm) of litter and adding more later because the litter will not be deep enough to retain any heat and will soon become cold and damp.

Dry warm litter is essential if the good bacteria are to multiply and assist in keeping the litter friable. Make sure the drinkers are set at the right height and the water level in the bowl is not too deep, so that if the birds knock the bowl they do not spill water all over the litter.

A plastic roosting pit floor.

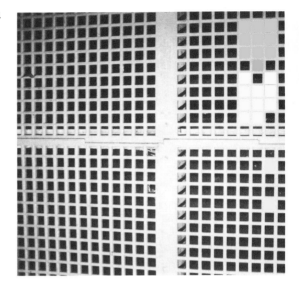

Roosting Pit Space

The raised roosting pit area should be large enough for all the birds in the house to be able to sleep side-by-side on it at night. This allows the litter area to dry out, as about 33 per cent of the droppings will fall into the pit. The worst systems of deep litter housing have only a few wooden perches standing in the deep litter with no roosting pit at all, which means that the litter beneath the perches soon becomes soiled, encouraging bad bacteria to grow, and the birds' dirty feet contaminate many of the eggs, reducing them to second quality. The best deep litter systems will have 33 per cent roosting pit area with 67 per cent deep litter area. The larger the slatted floor area is in relation to the deep litter area, the greater the chances are of eggs being laid on the litter or slats and not in the nest boxes. With 33 per cent roosting pit area, birds stocked at 12 per sq. metre (10 birds per sq. yard) will be able to sleep comfortably on the roosting pit at night. Some perches standing on the slats will be required, depending on the stocking density in the house (*see* pages 106–107).

Nest Boxes

The nest boxes should be situated on top of the roosting pit, running along the whole length of the egg collecting passageway, and can

74

be made of either metal or wood. The advantage of wood is that it is warm and natural to the birds, and there is no doubt that they prefer wooden nesting boxes when given a choice. The slight disadvantage with wooden nest boxes is red mite, which like to live between the joints, but this will only be a problem if you do not spray on a regular basis to keep them under control. Red mite do not like metal nest boxes. It is advisable that the light in which the nest boxes stand is mellow, not bright.

There are three types of nest box in use today, the communal, individual and semi-communal.

Individual Nest Boxes
These nest boxes are usually two, three or four tiers high and 2.7yd (2.4m) long and can be wooden or metal. They have roll-away egg floors and can have either manual or automatic egg collection. Each single tier is divided into eight or nine individual nest boxes and each nest will be sufficient to accommodate up to six birds. This type of nest box can be adapted for every situation: front or rear

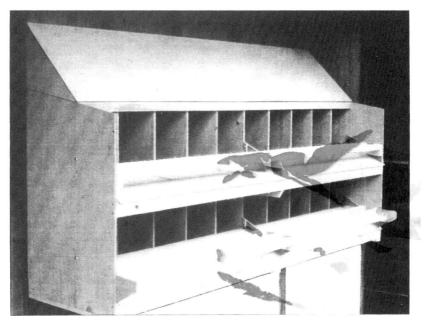

Individual hole nest box, with partitions to stop perch running, create a cosy corner.

egg roll-away and centre or side egg collecting passageways or no passageway at all. They can be equipped with folding flight rails, which the birds can use to get into the nesting hole, or which, folded up, will keep the birds out of the nest box at night. The roll-away floors are made of wire and covered with a plastic mat. An egg cushioning strip is fitted to bring the eggs gently to a stop at the end of the egg collecting cradle and manure deflectors reduce any manure from one tier rolling out into the egg trays of the tiers below. This type of nest box can be used on large or small egg producing units and is a good all-round nest box.

Semi-Communal Nest Box

This is one of the most successful nest box designs to be used in modern free range, barn or perchery commercial egg producing units. It is almost identical to the individual nest box but the nesting holes are nearly double the size. Although the same length, 2.7yd (2.4m), because the nesting holes are larger it is possible to increase the number of birds per nest by 25 per cent. Both the individual and the semi-communal nest boxes can be fitted with partitions to

Semi-communal nest box, again with partitions.

reduce the incidence of 'perch running' (*see* Glossary) and to provide birds with another 'corner' nest box in which they love to lay, thus reducing the chances of floor eggs in other undesirable parts of the house.

Communal Nest Box
This is a box 72in long by 24in deep (183 x 61cm) with a single entrance hole in front and a sloping roof which is hinged on top for easy egg collection. The dark warm conditions inside the box make an ideal breeding place for fleas and red mite. This box will provide a nesting place for up to 70 birds and is only used on small-scale poultry farms.

Automatic or Manual Nest Boxes
Automatic egg collecting nest boxes are expensive to purchase, usually at least twice the cost of manual boxes. Manual egg collection nest boxes, in a well-designed poultry house, are quick and easy to operate and there is less to go wrong. The beginner is unlikely to be able to afford automatic collecting nest boxes and it is difficult to see how they would benefit a small poultry farm or house with less than 10,000 birds. Automatic nest boxes are suitable for large houses and large multi-house farms where costs could be reduced by employing fewer egg collectors.

Automatic egg collection nest box.

*A central egg collecting
passage.*

*A side egg collecting
passageway.*

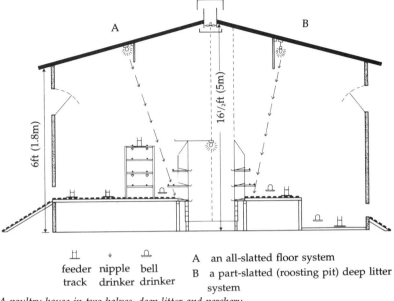

feeder track | nipple drinker | bell drinker

A   an all-slatted floor system
B   a part-slatted (roosting pit) deep litter system

A poultry house in two halves, deep litter and perchery.

THE ALL SLATTED FLOOR SYSTEM

This is one of the most popular and successful free range commercial egg producing systems in use today and has many advantages over the deep litter systems:

1. Stocking densities can be much higher, thus reducing overheads and daily running costs. The extra heat generated will also reduce feeding costs as the birds will not be eating feed to keep warm.
2. With no litter on the floor for the birds to lay their eggs on, they are more inclined to use the wood and plastic roll-away nest boxes, reducing the chances of floor eggs.
3. There is no need to teach the birds to use the roosting pit.
4. With no litter in the house there is a big reduction in dust levels which is good for the birds and also for the staff.

Scratching and Dust Bathing Areas

Scratching areas can be added to the all slatted floor system but the birds must not have access to them until they have learned to

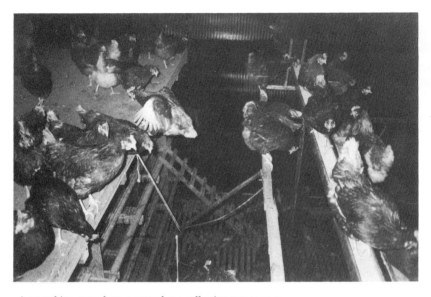

A scratching area above a central egg collection passageway.

lay in the nest boxes at 30 weeks of age. Even at this age they will still lay some eggs in the scratching area if they have access to it during the laying period. The only way to prevent this is to open it up in the early afternoon.

The best material to use in the scratching area is 2½in (6.5cm) of sand, as the birds cannot scratch it out and it is very absorbent. One of the best places to position a scratching area is above the egg collecting passage as this is easy to screen off from the birds, but ensure that the lighting in this area is not darker than the rest of the feeding area or they will certainly lay their eggs up there.

There is no doubt that birds do enjoy using a scratching area.

Nest Boxes and Other Equipment

The nest boxes for an all slatted floor system are the same as those recommended for the deep litter system. The boxes in both systems must be placed on the slatted floor so that the nesting area of the box is over the deep pit, not over the egg collecting passage. This will prevent any droppings falling in the egg collecting passage (*see* opposite). Feeders, drinkers and perching requirements for the all slatted floor system can be found in Chapter 5.

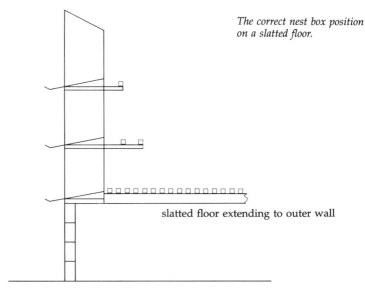

The correct nest box position on a slatted floor.

slatted floor extending to outer wall

THE ARRIVAL OF NEW BIRDS

Ensure that the birds are divided up evenly into the pens and that the total number of birds delivered is correct by counting the birds off the lorry into the house. Do not have too many birds delivered in one day and make sure there are plenty of people to carry the birds in with care. Reject any very small birds. In all slatted systems put some of the birds on the perches as this will encourage other birds to perch. Birds going into a deep litter system must all be placed on the roosting pit. The house should be ready for the birds and all the equipment checked and set up as follows.

Feeders

Flat Chain Type
Set the automatic time switch to run the feeder round the house five times each day, starting thirty minutes after the lights come on. Try to arrange the feeder so that the track passes along the front of the nest boxes first, as this helps to encourage more birds over to the nest boxes.

Overhead Tube or Pan Type
Ensure all the tubes or pans are full of feed.

81

Ventilation Controls

Set thermostats to maintain the temperature at 64°F (18°C). The minimum air flow should be set at 10 per cent of the maximum air flow. This could be less in the winter, but it is important that the fans do not go off altogether.

Water

Add vitamins to the water for the first three days and check that all the drinkers are set correctly and that they are not overflowing. It does help to encourage birds to go near the nest boxes if some drinkers are placed between the feeders and the nest boxes. This is because the birds' natural routine is to eat, drink and then go in the box to lay.

Lighting and Luminosity

Check how many hours' light the birds have been receiving during the last weeks of rearing and set the time switches to give the correct amount of light, bearing in mind the lighting programme being used. Remember never to reduce the light hours at this or any other stage of the birds' laying life. Make sure the luminosity of the laying house is brighter than the luminosity of the rearing house, but within the constraints suggested on page 35.

Electric Fence

Check that the electric fence inside the house is working properly and, if possible, leave it on as the birds are put in, otherwise they might break the fence down before they have all been unloaded.

THE MANAGEMENT ROUTINE

The management of all the systems is almost identical, except for a few minor differences where deep litter is involved, for example the birds have to be put on the roosting pit.

Day One

On the day of delivery, in deep litter systems every bird must be placed on the roosting pit and remain there until the second day,

Use plenty of people to unload the birds.

Prevent any accidental internal injuries to the egg laying organs of the birds by keeping two fingers between the birds' legs whenever you handle them.

and in all slatted floor systems some of the birds must be placed on the perches. Shut up all the nest boxes to prevent any birds perching in them at night; this is the first step towards keeping the nest boxes clean and reducing the number of dirty eggs. When all the birds are settled in the house, turn the lights off for a minute and, when the lights are back on, run the feeder for two minutes. This will get the birds used to the equipment suddenly going on and off. In the deep litter system make sure that all the birds are up on the pit half an hour before the main or roosting pit lights go off. In the all slatted floor system not all the birds will use the perches on the first night, as they would not have had time to anticipate the lights going off. Before leaving the birds, check that none of the drinkers have been upset by the day's events and that the birds are not crowding in the corners.

Day Two to Twenty-One Weeks

On day two the birds will look a little loose-feathered or 'crow-headed', which is a natural reaction to all the stresses and strains of moving, but by the third day they will look tight-feathered and become more active. In deep litter, continue to put any birds not already there on the roosting pit just before the lights go out every night. The birds' second night in the house will be the worst for this task as there will be a lot more birds down on the floor as they begin to explore the house. Each night after this the number of birds on the floor will gradually decrease and after about ten evenings only a few will insist on staying on the litter.

During weeks twenty and twenty-one the birds will start laying a few eggs. At this stage open the top and middle tiers of nest boxes and put some straw (not hay as this might cause compacted gizzard) inside the boxes to encourage the birds into the nest. The bottom tier should remain shut to encourage the birds to lay in the other tiers, and should not be opened until the first day that the birds achieve 10 per cent egg production. The straw will need replacing each day until the bottom tier of nests is opened up.

From the moment the first egg is found, walk round the house every two hours picking up any floor eggs. Failure to pick up floor eggs regularly will only make any problems worse. As the floor eggs are found, put an egg in each nest box, marking them with a red pen to show they are old 'dummy' eggs. These dummy eggs will encourage the birds into the nest boxes to lay. Remove all the dummy eggs when straw is no longer put in the nest boxes.

The correct equipment layout to encourage the birds to lay in the nest boxes; from left to right: perching, feeding, drinking, and laying.

Any increases in the hours of light should, for the first two weeks, be added to the morning hours, because to increase the light in the evening will upset the birds' roosting timetable and make it difficult to get them on the roosting pit. When walking round the birds, try to use the same route, always walking in the same direction. The birds soon learn to anticipate your next move, which will help to keep them calm and quiet as you walk around the house.

Now the birds are in lay a regular, *never* irregular, working routine must be kept to. Chickens are creatures of habit and perform best when they are kept to a strict routine. When letting birds out, do not open the pop holes until most of the eggs have been laid in the nest boxes, which is normally three to four hours after the lights come on. Make the first floor egg collection twenty minutes after the lights come on as a few of the birds will have laid their eggs on the floor during the night. If these eggs are left on the floor for long periods, the other birds will presume they can lay their eggs on the floor and, before long, there is a serious floor egg problem.

Twenty-One Weeks to Peak Egg Production at Twenty-Eight Weeks

By twenty-one weeks of age the birds' egg production will be between 5 and 40 per cent and a routine pattern of collecting floor eggs should have been developed, starting 20 minutes after the lights come on, with further floor egg collections every two hours. Record the percentage of floor eggs every day and relate the amount directly to the percentage of eggs laid in the nest boxes, as this will show whether the percentage of floor eggs is increasing or decreasing. As peak egg production approaches, the number of floor eggs should begin to drop. When egg production reaches between 40 and 75 per cent it is not unusual to find the birds all trying to get in certain popular nest boxes. Some birds will smother if they are not removed from the overcrowded nest boxes, so ensure that somebody is always on the look-out. Do not waste time picking up eggs from the nest boxes if only a few have been laid. It is always better to be looking for floor eggs than picking up eggs that are safe and clean in the egg cradles. Start shutting up the nest boxes in late afternoon, reopening them after dark. This keeps the birds out of the nest boxes at night. Check for broody birds when shutting up the nest boxes, and remove them to the broody-coop.

By twenty-eight weeks of age most of the artificial light should have been added to the time switch, some of which should have been added to the evening time and the rest to the morning (*see* page 54). Always bear in mind, when adding to the birds' lighting programme, that in the summer months there are seventeen hours of natural daylight, so avoid putting the lights on any earlier than 5 a.m. or the birds will have a sudden reduction of light hours as the winter months approach. Two or three extra feeds should be added by the time the birds reach peak egg production. Make sure the feed track is never empty. The best time to check is just before the feeder is about to go round again: if the chain has no feed on it then another feed should be added.

Twenty-Eight Weeks to End of Lay

Once the birds start to lay, get into a strict routine of picking up floor eggs, looking for any mortality and collecting the nest eggs.

Do not make any changes to the birds' routine and try not to change the staff too often. Remember that the older the birds are in mid-winter, the more susceptible they are to drops in egg production caused by stress of any kind. When the birds are laying, changes in the staff on the farm will sometimes reduce egg production. However, if all the staff wear the same coloured overalls the birds do not notice the different people quite as much, and this can be useful at weekends when the staff are most likely to change.

During the middle, and more often than not at the end, of the birds' laying life, the eggs become too large and the shell quality deteriorates. The most common cause of this problem is usually related to high specification feed eaten in excess because of a sudden drop in temperature. If the birds are middle-aged or old when this happens (when egg quality is not usually at its best), shell quality will deteriorate further and an increase in the number of bloody eggs will occur. The bloody eggs are a result of the birds' vents being torn by the sudden increase in egg size and can lead to vent-pecking if the problem is not checked. Sometimes a change to warmer weather will overcome the problem, but if that does not help a gradual change to a lower specification diet, or an end of lay ration, will usually put an end to the problem.

Sweep and clean up every day so that the birds get used to different sounds, and never forget that a farm is part of a food chain so it must always be tidy and clean.

End of Lay Tasks

Catching the Birds
Remove any equipment from the house before starting. Using the dimmer, increase the brightness of the light in the house and drive the birds down to the end where they are going to be loaded. When enough birds have been penned up (not too many or they might suffocate) turn the lights down quickly to near darkness and the birds will sit still, making it easy for the catchers to pick them up and reducing the amount of stress caused to the birds to a minimum. Always catch the birds in the dark hours if possible. Wear mittens to protect your hands and catch the birds from the rear, keeping them breast-downwards on the floor until all the birds required are gathered and then lift them up. If the birds are held off the floor as they are gathered up they will be able to flap their wings and turn on the catcher, which will result in the catcher getting bitten or scratched and unnecessary struggling for the birds.

Old Hen Disposal and Buying New Birds

Hens usually reach the end of their commercial laying life at about 72 weeks. Book old hens out early, as some old hen processors pay a bonus for early booking. Prices paid for old hens will vary considerably.

Most of the feed companies offer financing facilities for purchasing new birds. The first repayment is not due until the birds are thirty weeks old, by which time a reasonable income from the eggs should be coming in. The other repayments are usually due at three-monthly intervals. If possible, give the pullet rearer plenty of notice regarding the exact time the replacement birds will be required.

Cleaning Out the Poultry House

All the equipment must be thoroughly dusted, washed and disinfected before the new birds come in. Number all the equipment removed from the house, as it will then be easy to return everything to its original position. Repair and service any worn or moving equipment, paying particular attention to fans and feeders. Always use good disinfectants and change them on a regular basis, using one type for one flock and a different one for the next flock.

Outside the House

Rain, frost, sunlight and rest are the best forms of disinfectant for the land outside the poultry house. Plough up and re-seed over-used areas of land. Never let birds have access to stagnant or muddy water.

Induced Moulting

Moulting is induced by reducing the hours of light (*see* pages 52–3). The object of moulting birds is to give them a short rest from laying eggs, which will prepare them for a second laying period of about five months, making a total laying period of seventeen to eighteen months. The moult will also reduce their body-weight by 9–18oz (250–500g), which is essential to achieve good post-moult egg production.

Egg production will normally cease by the seventh day and start again between the sixteenth and twentieth days. The egg production of a force-moulted flock will be about 75 per cent of the original flock's average, with the peak being reached by the ninth week. A flock that does not perform well in its first laying period should never be moulted.

An induced moult is usually started at sixty weeks of age, but can be altered by up to eight weeks to suit individual circumstances. A programme can be obtained from ADAS (*see* Useful Addresses). There are both advantages and disadvantages to be considered before deciding to force-moult birds.

Advantages
1. There are no new birds to pay for, which can help in a cash-flow crisis.
2. It can improve average egg quality.
3. There are lower average working capital requirements.
4. The financial margins are increased when egg prices are low.
5. You receive higher returns on invested capital.
6. Larger eggs are produced.
7. There is less house cleaning and bird handling.

Disadvantages
1. It can reduce the financial margins when the egg price is very high.
2. Eggs are often too large and do not fit in the universal egg trays.
3. Better management is required.
4. Mortality during the moult and first laying period can mean that the house is under-stocked, creating higher feed consumption to overcome lower house temperatures.
5. Egg quality can be very poor if the second laying period is too long.
6. The roosting or deep pit space might not be deep enough to take the extra droppings and feathers.
7. Some egg packers will not take eggs from moulted birds.

9

Disease Recognition

The purpose of this chapter is to describe, in the simplest terms, the visual and audible symptoms of diseases that occur in birds living in free range, barn and perchery systems so that the stockperson can obtain professional advice quickly, before the birds' health and viability are affected.

Many of the disease problems experienced on poultry farms are caused or aggravated by bad management. Much of the disease and mortality in free range, barn or perchery layers could be prevented if the cause was understood, the symptoms recognized and the correct treatment given at once. Unfortunately, on many farms

Although a neck moult is a natural event in a bird's life, especially in older birds nearing winter, it is often a sign of disease or stress, particularly in young birds.

Feather loss caused by disease and management problems. The feather loss usually starts around the preening (oil) gland.

symptoms of some of the most common diseases are not recognized early enough for the vets or advisors to be able to avoid large drops in egg production and high mortality occurring in the birds. It is therefore most important that the stockperson should be able to recognize the early symptoms of disease.

When checking birds, it is important to stand still for a minute or two so that they get used to your sudden appearance; remain quiet and the birds will soon relax and show how they really feel. Always check for any respiratory symptoms at night, when the lights are off, listening for any coughing or sneezing.

SYMPTOMS OF ILL HEALTH

Early symptoms of ill health are usually difficult to detect in a flock of birds and can often go unnoticed. For example, in birds vaccinated against infectious bronchitis (IB) but which are suffering from a field challenge the symptoms are never obvious, in fact the birds soon begin to look better, but with careful observation the symptoms can be spotted and early action taken to reduce the negative effects on egg size, shell quality and production.

A bird showing early signs of sickness.

A very sick bird.

A young bird showing symptoms of disease.

The early symptoms of developing ill health in a bird, and the less obvious symptoms shown by vaccinated birds undergoing a challenge from a virus, are:

1. A flat back.
2. A level or dropped tail.
3. Head tucked loosely into the chest (crop).
4. Eyes partially or fully closed.
5. Head shaking (much more than they naturally do).
6. A slight loss of weight from the breast muscle, although this is not always evident.

Birds showing symptoms of being very sick will often be suffering from a mechanical breakdown of their internal laying organs, for which there is nothing that can be done to save them. Birds that are very ill will display the following symptoms:

1. A flat back.
2. A flat or drooping tail.
3. Head buried deep in the chest (crop).
4. Eyes completely closed for long periods.
5. Loose-feathered in appearance.
6. Sometimes one or both wings half or fully dropped.

93

7. Loss of muscle from the breast. Concave feeling when examining the muscle each side of the breast bone.

8. Audible and visual signs of respiratory disease: head shaking, sneezing, or coughing. Sometimes the birds try to relieve the irritation in their eyes by rubbing their heads on their neck feathers, leaving a wet ring around the middle of the neck.

9. Slow to move out of the way when someone is walking through the birds.

All symptoms that are noticed in the birds must be reported to the vet, especially if associated with a drop in egg production, egg size or a rise in mortality.

INFECTIOUS BRONCHITIS

IB is a viral infection causing depressed egg production and poor shell quality. Most of the laying flocks in the British Isles are vaccinated against IB, so the acute form of the disease is rarely seen. Nevertheless, there is no doubt that the virus is responsible for many of the drops in egg production which occur in many of the laying flocks. Both dead and live vaccines give good protection against the most virulent strain of the disease, but there are one or two variant strains, not completely covered by the vaccines, that are known to cause egg production drops and poor shell quality.

Regular re-vaccination of the birds throughout lay reduces the effects the virus has on egg quality and production, and it is therefore worth asking the advice of the vet or makers of the vaccine about how and when re-vaccination of the birds should be carried out.

Farms with more than one flock of birds (multi-aged) are particularly prone to irregular drops in egg quality and numbers, sometimes due to an IB virus moving round the different flocks. In birds which have been vaccinated properly, it will be difficult to spot any difference in behaviour when they are suffering from IB. The first change to be noticed, if the eggs are weighed weekly on the farm, is a slight drop in average egg size. The next change, three to five days later, will be a slight deterioration in egg shell quality. Sometimes, but not always, there will be a small increase in the mortality approximately ten days after the loss of egg quality. If the deterioration in egg quality and egg numbers is significant, multi-vitamins and medication can be beneficial. Always seek the advice of a vet.

FOWL PEST (NEWCASTLE DISEASE)

This is a highly infectious disease which can affect all commercial poultry, although the acute form is not seen very often as the majority of poultry are vaccinated against it. Fowl pest is a notifiable disease, which means that when a case is confirmed by a vet, the Ministry of Agriculture, Fisheries and Food (MAFF) must be notified. Producers outside Great Britain should check their local regulations regarding notifiable diseases. Symptoms of the disease include respiratory and nervous disorders, diarrhoea and, in the worst cases, a very high mortality. The best way to protect the birds is to make sure they are fully vaccinated.

INFECTIOUS BURSAL DISEASE (GUMBORO)

This disease is highly infectious and affects chicks from two weeks to six weeks of age, although some outbreaks of the disease have occurred in point of lay pullets. This disease damages the birds' immune system and the mortality can be very high. The damage to the immune systems of surviving birds can lessen the protection offered by any vaccination the birds have been given. Control of the disease is by live and dead vaccines.

AVIAN INFLUENZA (FOWL PLAGUE)

The severity of this disease will depend upon the strain of virus affecting the birds. There is a wide variety of symptoms associated with avian influenza, including coughing, sneezing and very poor egg production. The most virulent form of the disease, known as fowl plague, is notifiable and will cause high mortality and severely decreased egg production, but is rarely seen. There is no treatment or vaccination.

INFECTIOUS AVIAN ENCEPHALOMYELITIS (EPIDEMIC TREMOR)

This virus affects young chicks up to five weeks of age. The affected chicks show nervous symptoms and the mortality can be very high. The disease can also affect laying birds, but affected layers show no symptoms at all; the sudden reduction in egg production and the equally sudden return to full production, however, is a good indication that the epidemic tremor virus has infected the flock. Egg production can be reduced by up to 70 per cent over ten days and will usually return to normal about ten days later. Egg size will also be reduced during the infectious period. All replacement pullets should be vaccinated between nine and fourteen weeks of age. No treatment is available.

INFECTIOUS LARYNGO TRACHEITIS (ILT)

A highly infectious viral disease which affects replacement birds and layers, there are two forms of ILT, the acute and the chronic. In the acute form the birds will cough up blood and mortality will be high. In the chronic form the birds will cough and sneeze, but the mortality will not be as high. In both cases the birds can be seen stretching their necks out in an attempt to breathe. ILT can be spread by mechanical carriage (man, poultry crates, machinery, and so on) and once on a farm will continue to infect all future flocks. Vaccination is effective, but needs to be continued to protect every bird introduced to the infected farm. Medication will help the birds to recover. Seek the advice of a vet.

FOWL POX

This occurs in two forms. In the first, wart-like growths can be seen on the comb and wattles. The second form affects the inside of the mouth and trachea (windpipe). Young birds should be vaccinated if they are in an area where fowl pox is endemic. Control of the disease is by vaccination.

ENCEPHALOMALACIA (CRAZY CHICK DISEASE)

Usually seen in young chicks, this disease is caused by a deficiency of vitamin E, and leads to brain damage and a variety of nervous symptoms. The vitamin E deficiency can be a result of the bird's failure to absorb and utilize the vitamin or can occur because the vitamin has not been included in the feed at the correct level.

ADENOVIRUS ('EDS/76') OR EGG DROP SYNDROME

There are several different adenoviruses and EDS (egg drop syndrome) is one of them. This particular adenovirus will cause a loss of shell quality and colour. Many eggs will have soft shells or no shells at all, and many will be laid on the floor because the birds are too weak to reach the nest boxes. Diarrhoea and a poor appetite are sometimes evident. Vaccination gives the birds good protection. There is no cure, but medication can help to speed recovery. A vet will advise on the best treatment.

MAREKS DISEASE

Mareks disease is caused by a virus that affects the nervous system and also produces tumours in any of the internal organs of the body, and day-old chicks are most susceptible to infection. Although vaccination is carried out at day-old, the vaccine takes time to build up immunity in the chicks, so it is essential that the brooding house is completely clean. Usually symptoms are first seen in birds of three to five months old. When the birds are transferred from the rearing quarters to the laying quarters the stress of moving can sometimes activate the disease and a few birds will die ten to fourteen days after the move, slowly wasting away from lack of food and water. Mortality can continue throughout the laying period and losses of 30 per cent or more are not unusual in unvaccinated birds. There is no cure, but vaccination gives very good protection, provided the brooding house, in which the birds start, is clean.

PULLET DISEASE (BLUECOMB DISEASE OR NEW WHEAT DISEASE)

Young growers or pullets in early lay are particularly inclined to succumb to this disease, hence the name pullet disease. In birds which do go down with the disease the comb tends to look blue, which explains the second name. The third name is derived from the fact that outbreaks are more common in the late summer, soon after the new wheat has been harvested and used in the layers' feed. It was thought at one time that the new harvest of wheat actually caused the disease, but the real cause is unknown; fortunately it is uncommon. Affected birds will be thirsty and lose their appetite, but a foul-smelling whitish diarrhoea and a blue-purple comb are the more obvious symptoms. Egg production slumps may occur and mortality can be as high as 20 per cent if the early symptoms go unrecognized. Medication of the drinking water will successfully treat most cases.

PASTEURELLOSIS (FOWL CHOLERA)

Caused by a bacterium, mortality as high as 90 per cent can result from an outbreak in a laying flock. The symptoms of pasteurellosis vary widely: in the acute form there can be a sudden rise in mortality of apparently healthy birds; in the chronic form birds will discharge a profuse greenish-yellow diarrhoea and have difficulty in breathing. High mortality will sometimes continue until the end of lay unless some form of medication is given. Rats and mice can re-infect subsequent flocks of birds so regular baiting is essential. Vaccines are available.

COCCIDIOSIS

There are nine species of coccidia affecting domestic poultry flocks: *Eimeria tenella, E. necatrix, E. maxima, E. brunetti, E. hagani, E. praecox, E. mitis, E. acervulina* and *E. mivati.* Each of these species infects different parts of the birds' intestinal tract. The coccidia, small protozoan parasites, are passed in the form of oocysts or eggs from the intestine of an infected bird in its droppings on to the floor, where they are picked up by susceptible birds which will, in time, themselves develop coccidiosis. The symptoms of the disease will vary according to the species infecting the bird, but in general you can expect a drop in feed consumption, rapid loss of weight, birds huddling together, ruffled feathers, and diarrhoea sometimes tinged with blood. Mortality will vary according to the species of coccidia, which can occur at any age, although incidence is low in old birds. Drugs will provide the birds with an effective treatment. Vaccines are available, but are expensive. Pullets reared on floor systems usually develop a strong natural resistance to the disease, which can last a lifetime. Cage-reared birds are not advisable for extensive systems.

AVIAN SALMONELLOSIS

Some salmonellae infections which adversely affect the health of poultry can be transmitted to man, sometimes causing illness and death, and their control is therefore an important public health requirement for the poultry. The two main types of salmonellae that can affect the health of man are typhimurium and enteritidis. In breeding birds the disease is controlled by blood testing and the slaughter of positive reactors.

The infection can occur at any age. In young birds the symptoms of the disease are loss of appetite, diarrhoea (pasty wet vents), birds huddling in groups (similar to the huddling caused by chilling), and mortality up to 50 per cent. Survivors can become carriers of the disease. In the adult birds the symptoms are not as obvious and mortality is variable. The birds will sometimes lose weight and show signs of weakness, and some flocks will suffer from diarrhoea and/or a drop in egg production. These symptoms are not, however, characteristic and in all cases where mortality is greater than usual bodies should be sent to a laboratory for examination. Medication and multi-vitamins in the water will reduce mortality.

SPOTTY LIVER SYNDROME

This disease usually affects young layers up to the age of thirty weeks. Apparently good birds drop dead inside and outside the house. The livers of the affected birds are covered with little white spots. After the initial outbreak of the disease, subsequent flocks usually become infected. Early diagnosis and treatment is very important if large egg production drops and high mortality are to be avoided. The cause of the disease is unknown, but early treatment with medication is successful.

MYCOTOXICOSIS

Mycotoxins are produced by fungi growing in raw materials or in the finished feeds. The effects of the toxins on the birds can vary, from birds looking slightly 'off-colour' to causing substantially high mortality. The main symptom is reduced appetite as the feed is unpalatable; also the birds will sometimes appear to be thirsty and a drop in egg production and egg size will occur in birds in full lay. Retarded maturity will occur in young pullets if the infected feed is their first layer's ration at sixteen to eighteen weeks. No treatment is available, but put the birds on multi-vitamins while the feed is being tested for mycotoxicosis.

AVIAN MYCOPLASMOSIS

There are two types that affect commercial laying hens: *Mycoplasma gallisepticum* and *Mycoplasma synoviae*.

Mycoplasma Gallisepticum

This form of the disease causes a minor respiratory infection in the birds, leading to a slight drop in egg production and shell quality. It remains with the birds throughout their lives and drugs can only control the infection, they cannot eliminate the disease. Blood tests are needed to confirm the disease is in the birds. *Mycoplasma gallisepticum* increases the severity of other diseases the birds might encounter.

Mycoplasma Synoviae

This is the other form of mycoplasmosis which can affect commercial laying birds, the symptoms of which are swollen joints and feet. It can be responsible for depressed egg production and can be controlled with medication but not eliminated. The disease can only be identified by blood testing.

AVIAN CLOSTRIDIAL INFECTIONS

Several species of clostridia cause poultry disease; one in particular, *Necrotic enteritis*, often affects free range layers. The various species of clostridia are widely distributed, particularly in the soil. Affected birds will be underweight, lethargic and have a 'tucked-up' appearance. Mortality can be as high as 10 per cent. Medication does control the disease, but to help avoid infection ensure the birds do not drink from dirty, stagnant puddles, especially pools or puddles in areas where there is a high concentration of birds, for example narrow passages taking birds to an adjoining grazing area. There is a high incidence in birds living in dirty houses and muddy stagnant pastures.

HISTOMONIASIS (BLACKHEAD)

A disease that is usually associated with turkeys, histomoniasis does sometimes occur in laying hens. Affected birds become weak and lethargic, and the feathers around the vent are often stained yellow because of the sulphur-yellow diarrhoea. Medication in the feed or water can provide the birds with effective prevention or treatment.

AVIAN COLIFORM INFECTIONS, COLI-BACILLOSIS

Usually a secondary infection which invades poultry of all ages, *E. coli* sometimes affects layers after other infections, such as infectious bronchitis, have challenged them. Poor ventilation can also create the ideal conditions which encourage *E. coli* to infect the birds. The main symptoms are listlessness and a 'tucked-up', loose-feathered appearance. Medication provides an effective treatment.

VENT GLEET

An inflammation of the cloaca with a highly offensive odour, this disease is rare nowadays. Similar symptoms are found in birds which have been vent-pecked by other birds usually because of bad management, for example incorrect lighting or insufficient feeders or drinkers. There is no effective treatment, other than checking that the management is correct.

MITES

These are blood-sucking parasites which invade the birds, leaving them anaemic and emaciated. The red mite causes great irritation as it feeds on the birds' blood while they try to sleep at night. Heavy infestation by red mites will reduce egg production and some birds will eventually die. The mites are about 1mm long, light grey or bluish-grey in colour, but red when gorged with the birds' blood, depending on the stage of digestion. During the day they hide in any crack or crevice they can find. Check for red mite by running fingers over the surface of the nest boxes and woodwork, especially where it is dark. When the egg collectors complain about an irritation on their hands and arms because of red mite, you can be sure that the hens are really suffering. Their range of colours from grey to red makes them easy to identify and they are easily destroyed by spraying one of the many multi-purpose sprays which will kill red mite, lice and northern mite. Northern mite live continuously on the bird, but are rarely seen today.

The body louse is the commonest of all the louse species. It is about 1–2mm long and is pale yellow in colour. It is found mainly around the vent and tail; when the feathers are parted lice may be seen over the surface of the skin. The eggs or nits are found in clusters at the base of the feathers. Treatment is with a multi-purpose spray available from agricultural stores or vets. A repeat treatment is usually necessary. Keep all houses clean, tidy and well ventilated to reduce the risk of a red mite or louse infectation; the louse is rarely seen on a clean farm.

CONSTITUTIONAL DISORDERS

It is important to try to recognize birds that have died from 'wear and tear'. Conditions of the egg laying tract, for example, are common as this is a very active and sensitive part of the bird's anatomy, sometimes producing 300 eggs in 52 weeks. The following are some of the causes of 'wear-and-tear' deaths.

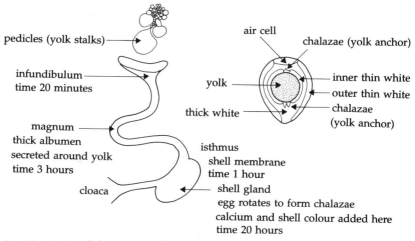

An active ovary and the structure of an egg.

Prolapse of the Oviduct

This is the protrusion of the oviduct through the vent and is more common in young layers, particularly in heavy layers. It is often caused by abnormally large eggs being forced through the vent.

Egg Bound

An egg bound laying bird is unable to pass an egg because the egg is abnormally large, there is a broken egg in the oviduct or the bird is in an over-fat condition. The abdomen of the bird will be full of yellow egg pieces, which sometimes give off an offensive smell if *E. coli* is present (usually as a secondary infection).

Gizzard Impaction

This affects chickens of all ages and is most common in late summer when birds have access to grass that is coarse and too long. The long coarse grass blocks the gizzard and the bird starves to death. Keep grass short, especially when young birds are going outside for the first time. Graze grass with sheep, if possible.

WORMS

A great variety of worms may infest poultry, especially birds on free range. Free range layers have contact with their own droppings and those of other birds within the flock, which makes them more susceptible to an infestation. Worms can have a number of effects on the chicken.

1. Reduced growth rate in young birds.
2. Reduced egg production.
3. Poor feed conversion (high daily feed intake).
4. Loss of condition.
5. Poor feathering.
6. Pale and weak shelled eggs (more seconds).
7. Increased susceptibility to other diseases.
8. Watery diarrhoea.

If a worm is seen in the droppings it does not mean that the birds are full of worms. Most flocks will have some birds with a few worms, but they should be treated before it causes a problem.

The worms normally found to cause problems in commercial layers are:

1. The large roundworm (*Ascaridia galli*), 1½–3in (38–76mm) long, white or yellow-white in colour.
2. Tapeworms (*Raillietina spp*) microscopic to 6in (15cm) long, white in colour. Small ones are difficult to detect.
3. The ceacal worm (*Heterakis gallinarum*) ¼–½in (6–12mm) long, greyish-white in colour, found in ceacal tubes.

Prevention

Ask the rearer to worm the birds on the rearing farm or worm them yourself before they come into lay. It is important to worm the birds twice. Birds which are force-moulted should be treated when they are out of lay. If it is known that a worm problem exists, do not scratch feed birds outside on the grass. Ask a vet for advice on the correct treatment.

— 10 —

Egg Producing on a Small Scale

Regardless of whether there are 6 laying hens or 6,000, the management of the birds will be similar. There are, of course, big differences in the size of housing required for smaller flocks, and there are many new poultry houses available which have been designed to keep between 30 and 100 birds in the modern perchery style and which can be used for free range production. Special feeding and drinking equipment can also be used for small flocks. Tube feeders can be used instead of automatic feeders and a hand fount can be used for water. The easiest method of feeding is to allow the birds access to a constant supply of ready-made dry mash. Do not use pellets as they are more expensive and make the birds fat. This method of feeding is also labour-saving as, if the larger type

Small house for small-scale egg production.

Larger type laying house for small-scale egg production.

of tube feeders and automatic bell type drinkers are used, the birds will not require feeding and watering every day. Small units can be equipped with roll-away nest boxes with outside or inside access. The house should be moved periodically to new pastures or be surrounded by four different grass runs, as this will give the birds access to fresh grass and allow the old run to be rested. Use an electric fence or chicken-wire 59in (150cm) high to keep the fox out and the chickens in. Provide artificial light during the winter months so that the birds get a minimum of sixteen hours of light per day. A low wattage night light (5 or 10 watts) will be sufficient to keep the birds in lay.

PERCHING REQUIREMENTS
AND POULTRY HOUSE CLASSIFICATION

It is necessary for the Ministry of Agriculture's Egg Inspectorate to assess a poultry house to classify it as being suitable for use as one of the following systems of commercial egg production. Outlined below are some of the requirements from EEC Regulation 1943/85. The full document is available from the Area Egg Inspector's office.

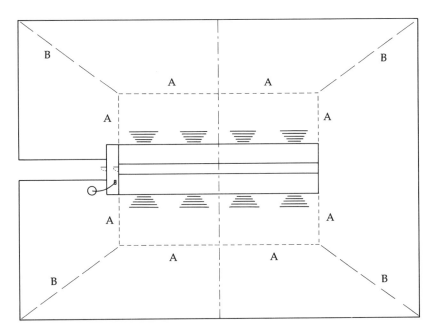

Alternate paddock grazing.

A   movable fence   B   permanent fence

1. *Deep litter*: the stocking density should not exceed 7 birds per sq. metre (5.8 birds per sq. yard); some provision should be made for the collection of droppings, i.e. a roosting pit of a sufficiently large area.

2. *Perchery or barn*: the stocking density should not exceed 25 birds per sq. metre (20.9 birds per sq. yard) in the space available to the birds and 6in (15cm) of perching space must be provided for each bird. This system usually has all slatted floors. It would be unwise to stock birds at 25 per sq. metre (20.9 per sq. yard).

3. *Semi-intensive*: the interior of the building must satisfy the conditions in 1 or 2 above. Maximum stocking density outside the building is 1,618 birds per acre (4,000 birds per hectare), although it would be unwise to stock land at this density. Eggs produced by this system can only be sold as semi-intensive and not as free range. The birds must have continuous access to open-air runs mainly covered with vegetation.

Perching and ventilation layout in a small-scale poultry house.

In all the systems mentioned above care must be taken to ensure that the whole of the floor area can be counted as floor space available to the birds. Obstructions, such as nest boxes, A-frames, H-frames or any other types of perching or nesting equipment, will have to be deducted from the total floor space available to the birds, unless they are at least 16in (40cm) above the floor level. The overall space taken up by any A- or H-frame without a droppings board underneath it cannot be included in the calculations. The area to be deducted is calculated as follows:

Perches in a vertical column − length of perch or longest perch in the column × 12in (30cm).

Perches in A-frame − widest section of frame + 12in (30cm) × length of frame.

Perforated platforms − widest section of platform + 12in (30cm) × length of platform.

Ask your Regional Egg Inspector to visit the poultry farm to advise on the correct amount of perching space required for the system of egg production preferred. Do this long before the house is completed, not when it is full of birds.

HOW TO REDUCE FLOOR EGGS

Following the guidelines listed below will help to reduce the problem.

1. The distance between nest boxes and the furthest point in the house (usually the pop holes) should be no greater than $6^{1}/_{2}$yd (6m).
2. Use brighter lights in the feeding area and darker lighting in the nesting area.
3. Stop stray light creeping through windows or cracks around pop holes. Spring- and summer-started flocks are often attracted away from the nesting area by light creeping in from places away from the nest boxes.
4. Put dummy eggs and straw in the nest. Keep replacing the straw.
5. Switch the electric fence on before putting the new birds in the house and check that there is a good earth.
6. Pick up floor eggs every one or two hours.
7. Avoid wide ledges and deep recesses in the pen which might make a comfortable nesting site.
8. Avoid shadows on the floor. Only the nest boxes should be in a mellow light.
9. Do not use clear electric lamps as they cast dark shadows.
10. Send the track feeder round in front of the nest boxes when it first starts its run round the house. This will entice more birds towards the nest boxes.
11. Put a few more drinkers near the nest boxes so that the majority of the birds will drink there after feeding (because the feed is dry) and will then find the nearby nest boxes a very convenient place to lay in.
12. When looking for floor eggs, always move in the same direction around the pen. The last movement you make must be one that moves the birds towards the nest boxes and not away from them.
13. Low feeders and drinkers (lower than the level of the birds' backs) will make good nesting places for floor egg layers. Low

feeders, especially track feeders, will make it difficult for the birds to see and reach the nest boxes, so keep feeders and drinkers at the right level.

BROODINESS

The broody factor in the genetic make-up of a bird is dominant to the recessive factor for non-broodiness. Therefore, in order to reduce the amount of broodiness in the modern hybrid layer the breeders eliminate all the birds in their breeding lines that show any signs of broodiness. Nevertheless, broodiness can still be a problem.

Broodiness is always more of a problem in young birds who start laying in the spring, as this is the natural time for all birds, not only chickens, to start sitting on eggs to incubate them. In the wild, after the bird has laid a clutch of 15 to 20 eggs she would normally go broody and would be encouraged by the warmer temperatures of the spring weather to sit on her eggs. Even though modern breeding techniques reduce the broodiness in the birds to a low level, as many as 8 per cent will still go broody in spring flocks

Removing the bird gently from the broody coop.

The broody coop.

if suspected broodies are not removed immediately from the nest boxes. As the last collection of eggs is made, check the nest boxes for any birds showing typical signs of broodiness, such as aggression when approached, birds who remain in the same nest box day after day, or any birds who have fluffed-up feathers and a dull comb. If the bird has been broody for a long time her breast will be bare of any feathers and will feel warm to the hand. She will cluck noisily, defying the poultry keeper to touch the eggs. All broodies must be put into a broody coop, which is a wire or slatted floored box in which birds can be placed to break them of their broodiness. Each bird should have its own coop, as this is the only way to be sure which bird is laying after the broody period. The longer a bird has been broody, the longer it will take to get her back into lay.

Broodies should always have the same lighting pattern as the rest of the flock; never reduce the hours of light to broodies or it will put them into a forced moult and they will be out of production for an unnecessarily long time. The broody birds must always have feed and water in the broody coop.

SPECIALITY BREEDS AND PRODUCTS

The commercial egg industry has, like many other industries, a value-added or specialist section of its market. These special products include very dark brown eggs, green eggs, white eggs and four grain eggs. Layers which produce four grain eggs are fed mainly on four different grains, which gives the eggs a special flavour. The bird that lays the dark brown eggs is called the Speckledy. These breeds are not available on the open market but are specific to Stonegate Farmers Ltd (UK) and can only be obtained from them through a special egg contract. Otherwise, with the obvious exception of four-grain layers, the management of special breeds is no different from that of standard breeds.

CULLING BIRDS

It is important that the stockperson on the farm knows how to cull a sick bird which is not worth keeping alive, either because it is unproductive or because it is suffering pain or discomfort.

Young Birds

In young birds aged up to seven days old the neck can be suitably broken by pressure with the thumb against a sharp edge, for example the edge of a door or table.

Older or Medium-Sized Birds

With the left hand hold the legs, whilst the right hand grasps the head with the palm against the back of the bird's head. With the head resting in the hollow formed by joining the forefinger and the thumb, bend the head upwards, using the thumb under the beak, and at the same time pull the head firmly forward, stretching the neck and dislocating the head. As soon as the separation is felt the stretching must stop; be careful not to pull the head off.

Culling a sick bird.

There will be some reflex action of the limbs, during which the base of the wings should be held to stop the bird flapping about.

Dislocation, if performed correctly, is instantaneous and humane and should be learned by all who care for animal welfare.

RODENT CONTROL

Rats and mice can carry all kinds of disease so, no matter how well the poultry house is cleaned at the end of the flock, rats and mice might soon re-infect the new birds with disease. Pasteurella and salmonellosis are only two of the many diseases rodents can re-infect the birds with, so regular baiting is essential if they are to be kept under control.

113

EGG QUALITY FAULTS

Meat Spot or Blood Spot

This applies to any small particle trapped in the white of the egg or a leakage of blood vessels around the ovum or in the oviduct, and is visible when the egg is candled. Spots can be caused by shock, stress, disease (IB) or thunder. There is no treatment, but give vitamins in the water for 48 hours.

Plum Egg

This is a variety of egg laid by some hybrids which is a mauve colour. Plum eggs are rejected by some egg packers and are best sold at the farm gate.

Double-Yolked Egg

This is the result of two yolks being released together from the ovary into the oviduct (infundibulum) and then included in one

Egg shell quality problems caused by IB, ND and stress-related problems. Banded eggs are produced by frightened birds or birds that are disturbed in afternoon and night rest periods.

Pale eggs are caused by disease, mainly respiratory, IB and ND or because the birds have severe feather loss.

shell. There is a higher incidence of double-yolked eggs in young birds starting to lay.

Watery Whites

Watery whites are caused by high ammonia levels in the poultry house or by an IB challenge interfering with the secretion of egg white from the oviduct (magnum), and is more often a problem in old birds. Check night ventilation for ammonia, and add multi-vitamins to the water for 48 hours. Medication of the water will help to solve the problem if it causes customer complaints.

Soft-Shelled and Shell-Less Eggs

This is a common fault in young layers just starting to lay, although old birds at the end of lay will also produce a few more soft-shelled or shell-less eggs. Young layers will grow out of the problem. For old layers, recent tests have shown that feeding ad lib oyster shell grit is highly beneficial. Contact your poultry specialist for advice.

Shell quality problems caused by disease and parasites: red mite, worms, IB, ND, ILT, EDs, MG, and MS. Loss of shell quality and shell colour sometimes combined with a loss in average egg size are indicators of health or management problems in a laying flock.

Misshapen Eggs

These can either have flat sides, body checks, rough ends, or be grooved, elongated, ribbed and have poor shells, pale shells or soft ends. It is not unusual for several of these faulty eggs to be laid every day, especially in old birds. An excessive amount of second-quality eggs, similar to those mentioned above, could be caused by a disease problem in the flock, IB, EDS, ILT, FP or mycoplasma.

Body-Checked Eggs

Body-checked eggs have a raised band around the centre of the shell and can be produced by nervous birds who react to slight noises, and so on. One of the major causes of this problem is too much activity in the laying house in the afternoon when the bird is 'manufacturing' the egg. Changes in staff at weekends or an irregular egg collecting routine can also cause the problem.

Blood-Stained Eggs

These are caused by sudden drops in temperature, triggering a large increase in the birds' feed intake, which pushes up egg size suddenly, tearing the vents of some of the birds, who are often over-fat. Try to keep a good average temperature in the poultry house. This problem can also be caused by cannibalism or vent pecking.

EGG PRODUCTION LEVELS

Drops in Production

A drop in egg production is the result of nature's protective mechanism. Basically, when a hen is in full egg production great demands are placed on her body and in many ways she is playing physiological brinkmanship. If something then arises that adversely affects her well-being, nature tends to protect the mother at the expense of the next generation, and the best way of achieving this is to stop egg production and allow the hen to worry about her own body. Disease, bad management and stress will all cause or contribute to a drop in egg production. Most of these causes could be avoided or the effect of them reduced by detection, prevention (vaccines) and good flock management.

Poultry Production Calculations

Hen Housed Production
Add up the total of eggs laid in the week and divide by 7. This will give the average daily egg production. Multiply this figure by 100 and divide the answer by the number of birds placed in the house at the start of lay. This gives the hen housed production percentage.

Hen Day Production
Add up the total of eggs laid in the week and divide by 7. This will give the average daily egg production. Multiply this figure by 100 and divide the answer by the amount of birds alive on the last day of the week. This gives the hen day production percentage.

Average Daily Feed Consumption
Divide the total amount of feed consumed by the birds by the

number of days the birds have been on the farm. This gives the total daily average feed intake for one day. Divide the total daily average feed intake by the number of birds living at the end of the week. This will give the average daily feed consumption for each hen.

Total Mortality Percentage
Multiply the total number of dead birds to date by 100 and divide by the number of birds started. This gives the mortality percentage.

Percentage of Graded Eggs
Grade the eggs from one day's total egg production. Multiply by 100 the number of eggs from each grade (separately) and divide by the total of eggs laid that day, including seconds. This will give a percentage for each grade which must include second-quality grade eggs.

Egg Mass
To work out average daily egg mass multiply the Hen Day Production percentage by the average egg weight for the week and divide by 100 to give egg mass in grammes. To work out weekly egg mass multiply the average number of eggs laid per bird for the week by the average egg weight. This will give the total egg mass produced in the week in grammes. To work out cumulative egg mass add up the egg mass figures for each week from the onset of lay up to the relevant week of production. This will give the cumulative egg mass, usually quoted in kilogrammes.

Glossary

ADAS Agricultural Development and Advisory Service.
Air Cell Found in the broad end of the egg. Provides air for the chick.
Air Convection The natural process of air movement which takes place as fresh cold (heavy) air enters the poultry house, falls and displaces the warm light (used) air which rises to the roof of the house to be extracted by fans or roof vents.
All-slatted system A slatted floor made of plastic, wood or wire over a manure collecting pit. (Alternative name for a deep pit system.)

Barn Eggs Eggs produced in a perchery type house or system where the birds do not have access to land or grazing.

Candling Visual examination of the shell and contents of eggs by holding them over a bright light.
Case of Eggs One box containing 360 eggs.
Chick The first six weeks of a bird's life.
Crow-Headed Term given to a bird with a crow-like head, or whose head feathers are loose or ruffled-up because of disease or stress, for example after moving birds to a new laying house. In the female it is often a symptom of a poor layer.

Deep Litter A system of keeping rearing, breeding and laying stock by covering all or part of the floor with white softwood shavings, cut straw or a mixture of both.
Deep Pit A system of collecting manure for the duration of the laying year. The birds walk on slatted frames which are made of wood, wire or plastic placed above the manure. The height the slats are placed above the manure will depend on the stocking density of the birds in the house, for example a pit one metre deep will be sufficient for a stocking density of 15 birds per sq. metre (12.5 birds per sq. yard) for one year's lay.

Egg Room Cooler An electronically powered fan cooler unit which is thermostatically controlled to keep the eggs cool and reduce the amount of moisture lost through the shell.

Floated-Up Tamping the wet chalk or concrete when constructing a new poultry house floor will bring to the surface a combination of moisture and small particles which leaves the surface smooth and easy to clean.

Free Range Egg An egg produced from birds living in a perchery or deep litter type of laying house who have access to fields, mainly covered with vegetation. The stocking density of the birds using the grazing area should not be in excess of 405 birds per acre (1,000 birds per hectare).

French Drain A trench 30in (75cm) deep × 12in (30cm) wide with a 4in (10cm) storm pipe laid in the bottom. The trench is filled with 2in (5cm) of beach stones.

Friable Description given to deep litter that is easily crumbled. Good deep litter is always friable.

Grower The young bird aged between six and sixteen weeks.

Light Meter A hand-held meter which will instantly display the brightness of light in any part of the poultry house. The light is usually measured in lux.

MAFF Ministry of Agriculture, Fisheries and Food.

Moult The shedding and renewing of feathers.

NFU The National Farmers' Union, which provides advice and insurance for farmers.

Packing Stations Established to collect or receive and market commercial eggs. Eggs are checked for quality (internally and externally) and graded into seven sizes before marketing.

Palatability How much food a bird will consume with satisfaction; associated with its liking for any food with which it is familiar. Meal with a high fibre content or which is fine or dusty becomes less palatable to the birds, sometimes reducing feed intake which can reduce the birds' ability to grow or lay properly.

Perchery A poultry house in which perches are essential because of the higher stocking density of the birds in the house, usually in excess of 7 birds per sq. metre (5.8 birds per sq. yard). A perchery can be used for barn egg or free range egg production.

Perch Running   The name given to birds who run along the perches to eat soft, cracked or broken eggs in the nesting boxes. Some perch runners will peck the soft shiny vents of the birds which have just laid an egg, often leading to cannibalism developing in the flock. Modern nest boxes have dividers between each set of nest boxes to help stop perch running and provide extra cosy corners for the birds to lay their eggs near.

Persistency   Associated with flocks of layers. Flocks showing good persistency will continue to lay well throughout the full twelve months of the laying year.

Pullet   A young bird older than sixteen weeks.

Roosting Pit   A raised roosting area within the poultry house which provides a place for the birds to eat, drink and sleep, with a collection place for droppings out of reach of the birds. This allows the litter to dry at night (deep litter systems only). The roosting pit can be made of wire, wood or plastic sections.

Semi-Intensive Egg Production   A system of producing eggs using a perchery or deep litter house where the birds have access to fields, mainly covered with vegetation, which are stocked between 405 and 4,000 birds an acre (1,000 and 10,000 birds per hectare). The eggs cannot be sold as free range.

Slatted Floor   Most perchery and barn egg producing systems use all slatted floors where the birds have no contact with the manure. Some all slatted systems provide a scratching area inside the house. The slats are made from wood, wire or plastic and are usually 1in (25mm) apart (wood) or form 1in (25mm) squares (wire and plastic). All slatted floor systems are sometimes referred to as deep pit systems.

Useful Addresses

Agricultural Development Advisory Service (ADAS)
Nobel House, 17 Smith Square, London SW1P 3JR
Department of Agriculture for Northern Ireland
Loughry College, Cookstown, County Tyrone BT80 9AA
Department of Agriculture & Fisheries for Scotland
Pentland House, 47 Robb's Loan, Edinburgh EH14 1TW
Domestic Fowl Trust
Honeybourne Pastures, Honeybourne, Worcestershire WR11 5QJ
Eurbrid
The Hatchery, South Raynham, Fakenham, Norfolk NR21 7HL
(suppliers of the Hisex Brown layer)
British Free Range Egg Producers' Association (BFREPA)
The Chairman, David Trix, LKL Poultry, 71 Castle Street, Salisbury,
Wiltshire SP1 3SP
Family Cook
Hurst Farm, Crawley Down, West Sussex
(suppliers of small houses, equipment and feed)
Grassington Rangers
Grassington Farm, Warren Lane, East Grinstead Road, North Chailey, East Sussex BN8 4HX
(pullet growing and rearers)
Harper Adams Agricultural College
Edgmond, Newport, Shropshire TF10 8NB
International Poultry Services
Green Road, Eye, Peterborough, Cambridgeshire PE6 7YP
(suppliers of the ISA Brown layer)
Ministry of Agriculture, Fisheries & Food (MAFF)
Egg Marketing Inspectorate Headquarters, Room 202, 10 Whitehall
Place East, London SW1A 2HH
Mick Dennett, Free Range Egg Production Consultant
17 East View Fields, Plumpton Green, East Sussex
National Farmers' Union
22 Long Acre, London WC2
Poultry World
Quadrant House, The Quadrant, Sutton, Surrey SM2 5AS
(monthly publication)

Ross Poultry
The Broadway, Woodhall Spa, Lincolnshire LN10 6PS
(suppliers of the Lohmann Brown layer)
Shaver Poultry Breeding Farms (GB) Ltd
Bawdeswell, Dereham, Norfolk NR20 4QH
(suppliers of the Shaver 579 layer)
Sappa
The Hatchery, Grove Lane, Stanton, Bury St Edmonds, Suffolk IP31
2BB
(suppliers of the Hy-Line layer)
Smith Associates
23 Montrose Avenue, Leamington Spa, Warwickshire CV32 7DS
(poultry courses)
Willow Agriculture
Willow Farm, Ridge Lane, King Stag, Sturminster Newton, Dorset
DT10 2AU
(suppliers of housing and equipment)

Index

Page numbers in italics denote illustrations